ROBOTS

BRINGING INTELLIGENT MACHINES TO LIFE?

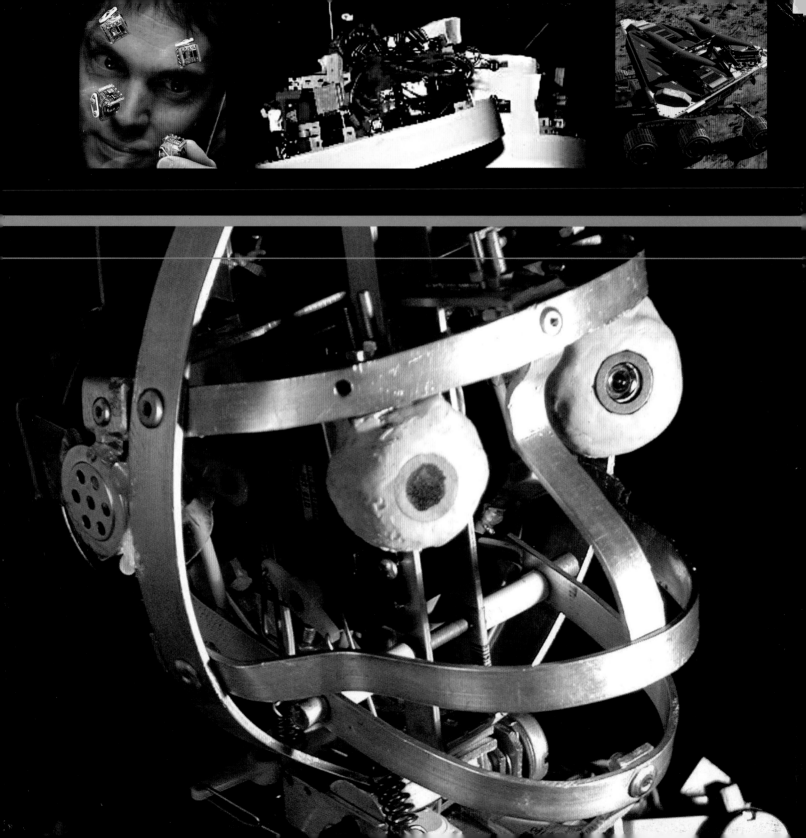

ROBOTS

BRINGING INTELLIGENT MACHINES TO LIFE?

CONTENTS

A QUARTO BOOK

First edition for the United States, Canada, and the Philippine Republic published by Barron's Educational Series, Inc., 2002.

Copyright © 2002 by Quarto Publishing plc

All inquiries should be addressed to:
Barron's Educational Series, Inc.
250 Wireless Boulevard
Hauppauge, NY 11788
http://www.barronseduc.com

International Standard Book Number
0-7641-5541-5

Library of Congress Catalog Card Number
2002105408

QUAR.ROBO

Conceived, designed and produced by
Quarto Publishing plc
The Old Brewery
6 Blundell Street
London N7 9BH

Project Editor Vicky Weber
Designer Elizabeth Healey
Copy Editor Gillian Kemp
Proofreader Alice Tyler

Illustrator Kang Kuo Chen,
John Kelly, Marshall Editions Developments Ltd
Photographer Martin Norris
Picture Research Image Select International
Indexer Diana Le Core

Art Director Moira Clinch
Publisher Piers Spence

Manufactured by Universal Graphics, Singapore

Printed by Midas Printing International Limited, China

9 8 7 6 5 4 3 2 1

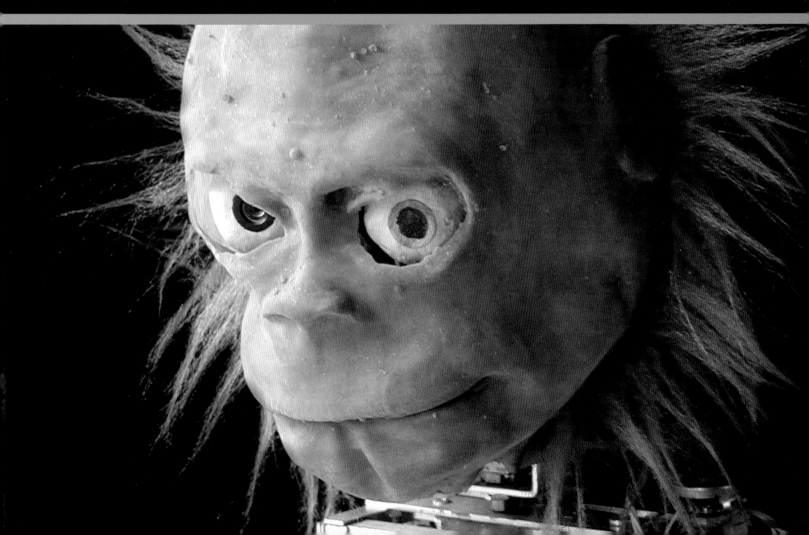

INTRODUCTION

This book is all about robots and the challenge of making them more functional and intelligent. But rather than seeing this as a mere engineering issue, the book looks at the lessons researchers are trying to learn from the natural world.

This trend started in the mid 1980s, when some researchers concluded that taking humans as the model for intelligence–as the field of Artificial Intelligence had done for 30 years–was not the best way forward. They began to look more closely at much humbler forms of life, such as insects, and were surprised by how clever these apparently simple creatures seemed to be.

While the science of biology can trace its origins back thousands of years, biomechanics–the study of how living things do what they do in engineering terms–is a very much more recent field of study. This new science is beginning to have an impact on robotics, especially in the area of research known as biomimetics–the attempt to copy the engineering of living things.

Whenever robots and intelligence are mentioned in the same sentence, someone invariably asks: "What if they become too intelligent?" Such fears have been around for as long as the dream of building an intelligent machine. But as we review the many problems that have to be overcome in pursuit of this goal, we'll see that even the simplest living things are still well ahead of the game. As they learn about the living world, researchers are more inclined to marvel at real animals than to feel they can surpass them, at least in the foreseeable future.

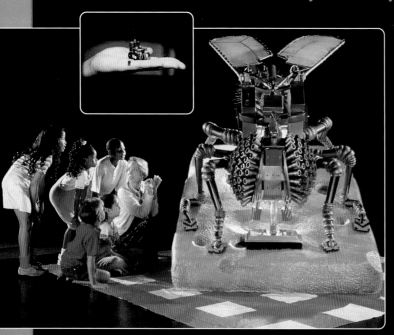

ABOVE

Experimental robots come in all shapes and sizes, from the tiny Ant robot that fits in the palm of the hand, to a robot fly as big as a family car.

The problems of mobility–on land, sea, or in the air–considered in Chapter 2 seem daunting enough, despite the many ingenious approaches robots take to getting around. The latest developments in sensing technology, without which the ability to move would surely lead to disaster, are reviewed in Chapter 3. But of course an intelligent robot needs to do a little more than move and sense, and Chapter 4 looks at the issues involved in thinking, deciding, and learning.

Later chapters focus on bodies, speculating on how robots could progress beyond the basic metal box with wheels and a computer; and on the issues raised by the need to interact with humans–the function and desirability of emotions, and the ability to socialize and make friends. In the final chapter we return to the fear of robot domination, examining how close robots are to becoming self-sufficient.

If reading this book makes you want to go off and build a robot of your own, then it will have served part of its purpose–the other part being to shed light on the facts about robotics and dispel some of the myths. University research groups and industrial laboratories are not the only venues for robotics research–the field is quite accessible to anyone with patience and some basic technical skills. At the end of the book you'll find pointers to kits, helpful books, and web resources. Perhaps when you look at these you'll feel–like some of those you will meet along the way–that you, too, want to give it a go.

RIGHT

How long before robots–like Robot 5 here from the movie *Shortcircuit*–become widely available in the commercial world? They might never overthrow their human masters, but they could very well soon escape the labs.

The dream of creating intelligent beings just like us goes back a long way—at least to the ancient Greeks about 2,500 years ago—as do fears of what might happen if we were to be successful in this task. The modern concern that robots might take over the world seems to be related to a misguided interpretation of Darwinian theory that has, for the most part, been universally accepted. Going back a bit further, we find transmuted fears of slave revolts, and before that, the association of science with magic, and the notion that challenging divine powers would inevitably bring punishment. Maybe human dreams and fears have always been intertwined like this.

Of course, to answer the question—What makes a robot "intelligent"?—it helps to have a clear understanding of what this word means. It is a fact that what we consider intelligent changes over time. 500 years ago, the ability to divide one long number by another—what we now call long division—was a

Chapter 1

THE QUEST FOR INTELLIGENCE

procedure known only to a small number of the intellectual elite. Now, however, we teach young children long division, and even a computer that can do millions of long divisions per second is not necessarily regarded as particularly intelligent. With animals or machines, it is difficult to judge whether an act of intelligence has occurred or whether it has merely been simulated. A monkey mimicking the actions of his trainer uses a degree of intelligence to interpret and reproduce the movements, but he cannot be said to be as intelligent as the trainer.

Many recent authors have challenged the view that human intelligence is a single thing, let alone measurable by a single number (the way IQ—or Intelligence Quotient—tests used to assume). The ability to carry out spatial reasoning, to write and speak fluently, to use the imagination, and to deploy emotional intelligence in social relationships have all been identified as capabilities present in varying amounts in different human individuals. Sufferers of autism sometimes possess exceptional intellectual ability, but we do not feel this makes them super-intelligent, nor would we design an intelligent machine to mimic autistic behavior. Then there are people who possess high intelligence but lack any kind of empathy—we call them psychopaths, and we definitely wouldn't want machines to behave as they do.

In this very human-centric view of the world we must ask: Are we really the only animal that could be called intelligent? Even very simple living things such as termites can produce amazingly complex behavior when they act together. Groups of lionesses hunting together have been observed to coordinate their behavior in a way that looks as if it has been planned, though we can assume it wasn't based on a quiet chat about what to do! Moving freely in the world, perceiving what is important, and interacting with others—all these are abilities we share with many other animals. Maybe an intelligent machine need not look or behave just like a human at all?

BELOW LEFT

R2D2 AND C3PO
Two very different robots from the *Star Wars* movie, but only one looks and behaves like a human.

BELOW RIGHT/BACKGROUND

INSECT ARCHITECTS
Social insects collaborate so that small, fairly insignificant living things are able to achieve what humans take a lot of technology to replicate. Building a termite hill does not require a "termite in charge," however.

FIRST DREAMS

From ideal servants to slaves in revolt, from gifts of the gods to products of human technology—the dream of creating a living being goes back thousands of years. But only when it became possible to combine elaborate metalwork with some source of power did robots finally enter the realm of the possible.

THE STORY STARTS WITH ART RATHER THAN TECHNOLOGY. A myth from Ancient Greece tells of the artist Pygmalion who sculpted an ivory statue of a woman so beautiful that he fell in love with it. He became so desperate to make his love real that he petitioned Aphrodite, the goddess of love, to breathe life into the statue. And, the story goes, she did. Life was a gift, and a mystery, of the gods. In a parallel story, Hephaestus (or Vulcan), an ugly lame smith and the closest thing the Greeks had to a god of technology, gave life to two female statues of pure gold. Their functions were to act as cupbearers on Mount Olympus, making them the first example of mechanical servants, a theme that runs through so many subsequent stories. The legendary Greek designer Daedalus—he whose son flew too close to the sun—was said by Aristotle to have created moving statues that guarded the entrance to the Labyrinth in Crete.

The Jewish tradition of the golem, or living statue, also reflects the notion that life is god-given. The making of a golem is described in the Talmud, and involved the chanting of combinations of letters from the Hebrew alphabet and the magical use of the "name of God." The most famous golem was said to have been made of clay in medieval times by the wise rabbi Judah Loew, the maharal of Prague (and a real historical figure). Life was given to it when the name of God was put on its forehead. Although it was created as a protector and helper of the oppressed Jewish community, some versions of the story—in a premonition of more modern fears—tell how the golem grew too powerful and self-willed, so that the name of God had to be removed from its forehead, returning it to its clay form.

MECHANICAL AUTOMATONS

By the eighteenth century, technology rather than divinity had become the preferred means of creating artificial life, as it remains to this day. The metalworking skills that made watches and clocks could be used to produce machines that moved—like the elaborate clocks in some German towns, with processions of mechanical figures that emerge when

the clock strikes the hour. Animated automatons, or manikins, carefully crafted to mimic human mannerisms and accomplishments, were popular in rich households. For example, Pierre and Henri-Louis Jacquet-Droz were famous for their elaborately lifelike mechanisms, which included a young woman who played the piano, and a boy who could draw and write.

The talented French engineer Jacques de Vaucanson (1709–82) was another master in this field in the mid 1730s. He approached the subject systematically and—because he wanted to use mechanical aids to illustrate an "anatomie mouvante" (moving anatomy), a three-dimensional atlas of human and animal organs—he began with a thorough study of anatomy. In pursuit of this goal, de Vaucanson built a mechanical duck. It could not only waddle in a ducklike manner, but could eat, digest fish, and excrete the remains in a natural way. The mechanism was driven by a weight and had more than 1,000 moving parts that were concealed both inside the duck and in the base upon which it stood.

WHAT ABOUT THE BRAINS?

The trouble with clockwork is that it requires winding up. It is hard to think of a machine as truly "alive" when it has a key sticking out of it. From Aristotle onward, it was supposed that a vital force was what made things "live" and that inanimate objects—from stones to machines—were lacking in this critical ingredient. What could substitute for the vital force that Aphrodite breathed into Pygmalion's statue?

Electricity was one answer. In the 1790s, Luigi Galvani linked a primitive battery to some frogs' legs and discovered that the muscles contracted, just as they would have done on a living frog. Perhaps this was the vital force. His nephew, Giovanni Aldini, proposed the use of electricity to revitalize the dead, though there is no record of anyone actually trying this.

Others took a different path. How could one get a machine to act autonomously—to carry out commands without human intervention? By the nineteenth century, real progress had been made. The Jacquard loom used punched cards to produce particular patterns in the cloth it wove. Babbage's differential engine design promised a machine that could do mathematical calculations. Meanwhile, steam power had been harnessed to provide machines that moved around, albeit under human control. It was all coming together…

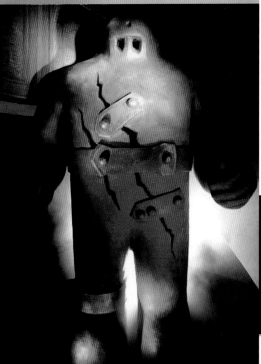

ABOVE RIGHT

STEAM ROBOT

Built in Canada in the nineteenth century, this robot was attached to a horizontal bar and could walk around in circles thanks to a ½ horsepower motor that drove jointed rods to move its legs. The cigar in its mouth functioned as an exhaust pipe.

RIGHT

LIVING DOLL

The golem, an inanimate clay model brought to life by the name of God, was a product of Jewish mysticism. A similar theme can be found in medieval alchemy, where the philosopher's stone was believed to have a life-giving force.

THEY'RE OUT TO GET US

Does our fascination with robots reflect our fears about our ability to control technology? Or is it that, as species evolve, we no longer have any guarantee that humans will remain at the top of the tree? Ever since Frankenstein, the theme of out-of-control robots threatening humanity has been a popular one. Even today, any scientist who says that robots will take over the world is guaranteed blanket media coverage for his claims.

LONG BEFORE ANYONE HAD ANY REAL idea about how to build an intelligent being, we were wracked with anxiety about what might happen if some-one did. Mary Shelley's most famous creation, from her 1818 novel *Frankenstein*, owed more to the tradition of the golem than to clockwork automatons, as the creature was assembled from a collection of human parts rather than machinery. Though the novel glosses over the matter of how the creature was actually brought to life, the film versions make it clear that it was electricity rather than God that produced the spark. The doctor pays a heavy penalty for his temerity: he finds that he has created not a man, but a monster. The creature, meanwhile, rejected by his creator, runs amok on a vengeful killing spree that ends in its own destruction.

ROSSUM'S UNIVERSAL ROBOTS

The word "robot," from the Czech term *robota,* meaning slave labor, was first coined by the writer Karel Capek in his 1920 play *R.U.R.* (Rossum's Universal Robots). The play concerns an industry in which humanoid servants are artificially created from biological material (like Frankenstein's monster) in order to serve the human race in factories and the military. Badly treated by their human masters, the robots revolt and kill nearly everyone. Because they are unable to reproduce, however, the robots are themselves doomed to die. All perish save two, a male and a female created by a last lonely scientist, and these wander off at the end of the play, hand in hand, to play robotic Adam and Eve. The theme of "revolting" robots is a persistent one, but in early stories it is the result of bad treatment rather than innate malevolence.

MAD SCIENTISTS

The idea that scientists might "play God" and let loose uncontrollable forces predates Mary Shelley by a long way. Even the biblical story of Adam and Eve, banished by God from the Garden of Eden, can be interpreted as a reflection of ancient fears about where too much knowledge might lead us. Medieval Christianity, which has had such a strong influence on Western European culture, was very suspicious of scientific enquiry. Only God could understand the workings of all things, and when humans presumed to share his knowledge, they committed the deadly sin of pride. The creation of the atom bomb seemed even to some of the scientists involved to be a case of releasing forces that might lead to human destruction, and today many people have grave concerns about genetic engineering. What if scientists create robots that could take over the world?

SURVIVAL OF THE FITTEST

The theme of robots rebelling against an unjust regime has obvious parallels with the history of human slavery. In the twentieth century, a new motive emerged as the process of evolution, first systematically explained by Darwin, received wide acceptance, albeit in a rather distorted form. Darwin argued that species evolved as the individuals best adapted to their environment were able to produce more offspring. Those that didn't adapt as well tended to die earlier with fewer or no offspring. The overall population would change over time, and a changing environment would mean that this process would never end.

This theory was misleadingly labeled "survival of the fittest" in the late nineteenth century, and developed into Social Darwinism, which suggested an actual struggle between the more fit and the less fit, rather than between animal and environment. It was certainly quoted widely to provide spurious intellectual backup for the distasteful ideas of racial superiority favored by various colonial regimes of the time. Could it be that, having used this idea to justify our own behavior in the past, we are now afraid that robots might one day act the same way toward us?

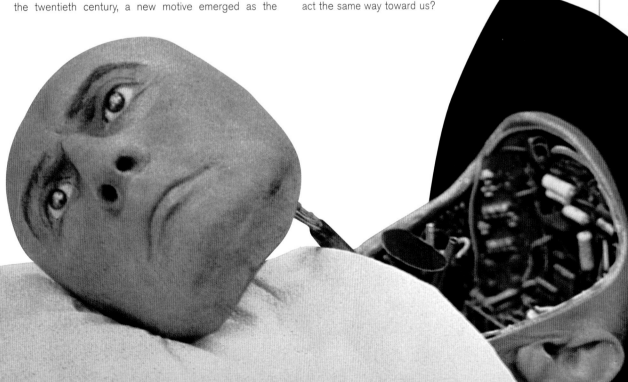

WHAT IS INTELLIGENCE?

An intelligent machine? What do we mean by "intelligent" anyway? Does it mean being able to play chess like a grand master? Scoring well on IQ tests? Does common sense play a part? Can only humans be intelligent, or will the growth of processing power produce intelligent computers?

AS SOON AS WE USE THE WORD "INTELLIGENT," QUESTIONS are raised. When the field of Artificial Intelligence (AI) officially emerged in the 1950s, Intelligence Quotient (IQ) tests were all the rage. A mixture of general knowledge questions and puzzles, these tests claimed to be able to determine a fixed IQ for any individual, which would objectively measure their level of intelligence. As a result, researchers in AI started to build computer programs that could solve the sort of puzzles found in IQ tests. The resoundingly titled General Problem Solver, created in the U.S. by Alan Newell and Herb Simon in the late 1950s, was such a program. The problems it solved were not those of world peace or global poverty, however, but the brainteasers that IQ tests were built on.

What about the ability to play chess? Surely a chess grandmaster must be very intelligent? If we choose to define intelligence in that way then we already have an intelligent machine. In May of 1987, the IBM computer Deep Blue took on the world chess champion Gary Kasparov and beat him. Ironically, although Deep Blue could work out its plays, it was unable to move the chess pieces. Chess is an example of a closed problem. It has a fixed number of pieces and clear rules on what is allowed, giving the computer the opportunity to consider more moves in a given time than a human. But is living in the real world a closed problem, with a finite number of moves?

These ideas of intelligence are all about pure thought, and do not involve any real interaction with the physical world—hence Deep Blue's problem with the pieces. We often think of intelligence as something the brain does on its own, without involving the rest of the body. In this scenario, the brain is thought of as being just like a computer.

LEFT

CONNECTED

In the 1920s, some people thought of the brain as a switchboard, in charge of routing thoughts and messages.

ABOVE

ENDGAME

Deep Blue beat the reigning world chess champion Gary Kasparov three games to two, with one drawn. Does this make it intelligent?

THE BRAIN AS A COMPUTER?

This metaphor is merely the latest in a succession that have been put forward throughout history. In the seventeenth century, thinkers were influenced by advances in plumbing, and thought of the brain as a hydraulic system. The philosopher René Descartes thought that the pineal gland acted on animal spirits in the liver to direct reasoning, so the brain worked like a tank and pumping system combined.

The industrial revolution brought with it a new set of metaphors. The brain was viewed as mechanical, containing levers, gears, and pulleys. The invention of the telephone changed the metaphor again and, during the 1920s, the brain became a switchboard. Of course, nobody has ever suggested that plumbing, gears, pulleys, levers, or telephones might produce intelligence—whatever we mean by the word.

That the computer metaphor seems to work a little better than these earlier ones might be because computers perform tasks that require thought when humans do them. And the rapid expansion of computer processing power has been impressive—roughly doubling every 18 months in what is known as Moore's Law, after the individual who first drew our attention to the fact. If the brain is like a computer, doesn't this mean that, as computers get faster and more powerful, they will inevitably become more like the brain?

THE BRAIN IS...

The brain is not a computer. We know that information in our heads is not held in static memory; we know that a neural network is not the same as a digital processor. Curiously, we also know that neurons work very slowly compared to a computer processor: their activity is measured in milliseconds, not nanoseconds. When humans carry out clever activities, like processing speech in real time, there is time for only about a hundred neurons to fire one after the other. It is the interconnection in parallel of thousands to hundreds of thousands of neurons that makes it all work. So the simple act of giving a computer more processing power will not make it more like the brain—it's what you do with it that counts. And what the brain does is still not completely understood.

AND INTELLIGENCE?

There is no definitive answer—even researchers in AI will only say that they are producing systems that "perform functions that require intelligence when performed by people," neatly dodging the question altogether. But how about this as a possible answer: Intelligence is "doing the right thing" in the real world. Not abstract thinking, not solving puzzles, but going shopping, putting the children to bed, driving to work. Tasks that Deep Blue would fail at completely.

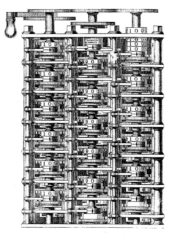

LEFT

HAND-CRANKED COMPUTER
In his differential engine, Babbage designed (but could not build) a real mechanical computer.

BELOW

SUPERBRAIN
Today's favorite metaphor is the brain as supercomputer.

LEFT

LIKE CLOCKWORK
One popular nineteenth century metaphor saw the brain as a complex mechanical device full of gears, levers, and pulleys.

GIVEN THAT WE CAN'T ROUTINELY LOOK under each other's skin to see what's going on, we have to rely on the way people behave to show us they are fellow humans rather than lifelike waxworks. So if a lifelike wax-work started behaving exactly like a human being, wouldn't we have to assume it was one?

One answer to this question can be summed up in the phrase: "A difference that makes no difference *is* no difference." It lies behind the first serious discussion of whether machines were capable of intelligence, featured in a paper by the brilliant British mathematician Alan Turing in 1950, at the dawn of the computer age.

Initially, Turing imagined a situation in which a human questioner could communicate with one man and one woman, without being able to see or hear them. The questioner could put any question—and the respondents were allowed to give misleading answers—in order to work out which was which. Let us imagine replacing one of the humans with a computer; again the questioner has to work out which is which. If it turns out to be just as hard to distinguish between human and computer, don't we have to accept that the computer is behaving intelligently?

OBJECTION!

The Turing Test turned out to be quite a tough one—no computer has performed nearly well enough to pass it so far. But there are a number of objections to it.

One is that what is going on internally also matters. A computer cannot know what the symbols it manipulates mean, so it literally has no idea what it is doing. So how can mere behavior be a valid test? To dramatize this argument, the U.S. philosopher John Searle invented an imaginary experiment known as the Chinese Room.

Suppose, he said, there was a large tent into which a Chinese speaker could post a message and get a mean-ingful reply. You'd surely say that the tent "understood" Chinese? But if you looked in the tent, you'd find a non-Chinese-speaking human operator who simply looked up the symbols in a table and found an appropriate output symbol, stringing them together to make the reply. This does not require the operator to know Chinese.

THE MEANING OF WORDS

However, both the Turing Test and the Chinese Room ignore another important issue. How is it that humans can

THE TURING TEST

As we have seen, there are many different ideas of what is meant by "intelligence." For Alan Turing, it was a command of language. The test he devised for machine intelligence is still debated today—is it fair, or even sensible? What other factors make us consider a living being intelligent?

understand language at all? The funny thing about a word is that neither the pattern of sound nor the symbol on the page bears any resemblance to the object it refers to. The word "moon" is not at all like the big white thing in the sky. So how do humans know what they are talking about?

One answer to this is that we attach our words—or strings of symbols—to the sensory experiences to which they correspond. The technical term for this is "symbol grounding," which means that words have to correspond to something in our real-world experience—for example, *seeing* the big white thing in the sky. This suggests that without symbol grounding there is no language. In other words, it is not the brain on its own that computes language, but the brain connected to the world in a real body, with real senses. In this case, neither a computer nor a tent would actually be able to handle language, and the Turing Test is doomed to fail. Unless we can construct a robot that interacts with the real world the way we do…

AND ARE WE ALONE?

Finally, all discussion of intelligence up to now has tended to assume that humans are the only intelligent animals. Do we really believe that? Many people think dolphins, dogs, and horses are intelligent. Even bees can tell other bees where to get honey. So perhaps it's back to the real world.

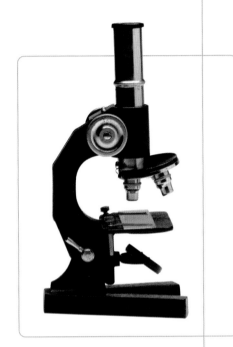
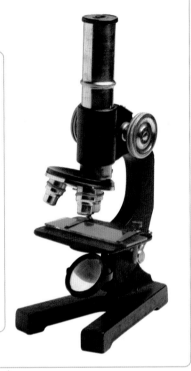

RIGHT

SPOT THE DIFFERENCE
You can probably tell that both these pictures depict the same microscope from different angles. So are you exhibiting intelligence?

Find Out More...
A great book is *Alan Turing—The Enigma* by Andrew Hodges (Walker and Company, New York) ISBN 0-8027-7580-2. Andrew Hodges also has an excellent Website at www.turing.org.uk/turing/.

Alan Turing's 1950 paper: "Computing machinery and intelligence" can be found on the Web at www.loebner.net/Prizef/TuringArticle.html; as can the Loebner Prize at www.loebner.net/Prizef/loebner-prize.html

THE LOEBNER PRIZE

In 1990, Hugh Loebner agreed with the Cambridge Center for Behavioral Studies, Massachusetts, to underwrite a contest designed to implement the Turing Test, pledging a grand prize of $100,000 and a gold medal for the first computer with responses indistinguishable from a human's. An annual prize of $2,000 and a bronze medal is awarded to the most "human" computer relative to other entries that year. So far, the grand prize remains unclaimed—entrants compete in an edited version of the Turing Test, where systems only converse on a single topic. In the inaugural competition in 1991, for example, some of the topics were "Whimsical Conversation," "Shakespeare," and "Romantic Affairs"—with eight computers and two humans. Five of the ten judges decided that "Whimsical Conversation" was done by a human, but it turned out to be the PC Therapist III from Thinking Software in Woodside, New York, running on a 386 PC.

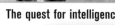

ANIMAL ANALOGIES

So far we have concentrated almost exclusively on humans, but if we want to make an intelligent machine that can live in the real world, shouldn't we look at other animals too? Focusing on what differentiates us from other species can blind us to all the things we have in common.

EVEN VERY PRIMITIVE FORMS OF LIFE, such as the microorganisms in pond water, incorporate mechanisms that allow them to exchange material with their environment. Homeostasis, for example—the process by which beings keep themselves in a functional state in a changing environment, gaining energy from sunlight or oxygen in their environment and producing waste—is common to all living things. In fact, elementary biology tells us that respiration, excretion, and reproduction are the definitions of what it means to be alive, from a single-celled organism to a human being. Oddly, in spite of the dream of a living robot, these functions have not been studied in great detail by roboticists.

But the single characteristic that divides living things into plants and animals is mobility—the ability to control one's location. Here, the most successful living organisms on Earth in terms of numbers are not humans at all, but insects. There are approximately 900,000 known species of insect—about three-quarters of all classified species in the animal kingdom. So in terms of living in the real world, insects do rather well. What's more, they are capable of some surprisingly clever things.

Insects have six legs, rather than four, like most mammals, or two, like humans. Having a larger number of legs actually makes mobility easier because it enables the insect to keep a number of legs on the ground at all times even when moving, thus avoiding problems of balance. But insect mobility means a great deal more than this. The common housefly can walk upside-down as well as fly. Other flying insects lift a relatively heavy body weight. The aerodynamics of bumblebee flight was a complete mystery for some time until close observation explained how they managed to lift off the ground. Then there are the grasshoppers, crickets, and fleas, all of which can jump vast distances in relation to their size.

Of course, mobility would be of little use to a living being if it could not sense the world around it—in fact, moving around without senses could lead to a rapid, and sticky, end. Insects are also interesting for the many ways in which they can sense the world. The common housefly, aside from its wonderful mobility, has eyes that give 360-degree surround vision. Crickets use song to locate their prospective mates. Other insects perform impressive feats of navigation, moving large distances from their base in search of food and finding their way home again. It is believed that some desert ants can sense polarized light from the sun for this task.

Social insects, such as bees, ants, and termites, show that even very simple living things can behave in complex ways. Termite mounds would be miracles of engineering if they were constructed on a human scale; they act as gas exchangers venting carbon dioxide from the nest and

BACKGROUND

PONDLIFE
Even microorganisms respire, excrete, and reproduce—the basic attributes of life.

BOTTOM

SOCIAL KILLER
This may look like survival of the fittest—but unless there is prey to be caught, the predator will starve.

RIGHT

BRAINY BEETLES
While no intellectual heavyweights, insects are surprisingly clever at moving and sensing.

allowing air in for ventilation. In spite of the extremes of temperature outside, inside temperatures vary by no more than a few degrees, and many termite species cultivate fungus as food.

As well as interacting socially with those of their own species, many animals have evolved complex relationships with other species, to their mutual benefit. The bacteria in your gut keep you healthy by maintaining an environment that is beneficial to them. Even predators like big cats have to live in balance with their prey, or risk future starvation.

All this interdependence—between animals and their environment, between animals and other animals—is basic to success, and making good choices is what allows the individual to play its part, to mate, and pass on its genes. From this perspective, the definition of intelligence seems very different, and moving, sensing, and interacting are the most important things. Taking this approach to robotics means trying to produce robots that can survive autonomously in real environments without falling over, running out of power, or crashing into their surroundings, while still producing interesting or useful behavior. These may not be the robots of science fiction, but they certainly represent a significant step on the path toward the goal of the living robot.

LEFT

BACKSCRATCHER
Birds keep the rhino clean and the rhino keeps predators away— a good example of species interdependence.

Mobility is vital for animals, and it is what separates us from plants. The ability to move under its own power is the defining characteristic of a robot—if it has to be radio-controlled, teleoperated, driven, or pushed, then it isn't strictly speaking a robot. But mobility is more than just part of a definition—it is essential to the way a robot interacts with its world and carries out its activity. If we think of intelligence as related to interaction rather than as deep but passive thinking, then the issue of mobility acts as a basic constraint on how intelligent a robot can actually be.

In order to move around, a robot needs some hardware to assist it—usually in the form of moving parts to which it can apply force. Engineers call these parts actuators, because they act on the environment. Actuators can take a variety of forms: wheels, legs, wings, or fins, to name a few. But actuators on their own are not enough—a robot also has to have some means of controlling what the actuators do if it is to move under its own power. Actuators plus a control system is what produces robot movement.

A key issue in getting actuators to control movement is the concept of "degrees of freedom." This is a basic mathematical idea that can be summed up by saying that there is one degree of freedom for every independent mode of motion. For example, a simple cylindrical robot moving around on a flat floor can independently move forward, or backward, and also rotate. This gives it two degrees of freedom. Now, give our robot the ability to fly—or to swim underwater—and it can do much more: it can move forward or backward along the three axes of three-dimensional space, and it can also rotate about each axis, giving it six degrees of freedom.

This isn't just true for robots, of course—humans also have degrees of freedom. However, because we

Chapter 2

MOVING MATTERS

are not simple cylinders skating around on the floor, we have quite a few more than six. Consider your hands. The top joint on each finger has one degree of freedom—it bends up and down. Five fingers, five degrees in all. Then the second joint on each finger also has one degree of freedom. The base joint has two degrees of freedom because it moves side-to-side as well as bending up and down. Then the joint in the ball of the thumb has three degrees of freedom. Add all of that up and we have twenty-three degrees of freedom, without even worrying about the wrist or the fact that most of us have two hands.

So why does this matter? Because each degree of freedom means independent motion, and a robot has to be able to control each one, as does a human. An intriguing thing about animals is that they manage to control many degrees of freedom very flexibly. Robot controllers built in the classic engineering style find it harder, mainly because each bit of control involves a set of equations; and for twenty-three degrees of freedom, that's a lot of calculation, even with modern computer processors.

No wonder roboticists have started studying how living things manage to move around so elegantly and flexibly. Ants scuttle around at high speed, considering their size, on six legs, but they don't seem to have enough physical bulk to hold the amount of processing power needed for all those degrees of freedom.

Few robots travel at great speed at present, unless we include military hardware, such as cruise missiles. Given that these move under their own control, we can think of them as a kind of robot. Still, a cruise missile only has to get to the right place and explode, which is not the most complicated or noble of interactions with the world. A galloping horse or a running human seems to be doing something much more complex and more beautiful.

BACKGROUND

MOVEMENT
Robots have to move under their own steam—the serpentinemotion of the snake is being seriously studied by some researchers.

RIGHT

ASIMO
Honda's 6-foot, 460-pound (1.8m, 208kg) P2 humanoid robot was developed over 10 years at a cost of around $100 million. Its successor, the Asimo, is the size of a large child and shares the P2's ability to walk like a human and go up and down stairs.

BOTS ON WHEELS

The easy way to have a robot move around is to put it on wheels—which is fine on nice flat floors or roads, but a bit trickier on rough terrain. So what about tracks? Now we can get over the bumps, but how about stairs? Sometimes wheels and tracks are very useful, and at least we know how to drive them; but robots using them don't always fit happily into human environments, where legs are the obvious way of getting around.

WHEELS, YOU MIGHT SAY, ARE THE OBVIOUS MEANS OF locomotion for a robot. With thousands of years experience of wheeled vehicles, we know how to make them, and the development of powered vehicles like cars has led to all kinds of useful engineering to help the power source control how the wheels move. Nearly all cheap robot kits, and even most of the available research robots, have wheels. Because they are simple and well understood, wheels are also often used on rovers for exploring the Lunar or Martian surface.

The most simple kind of wheeled robot has only two motorized wheels, with a free spinning wheel further back and halfway between, forming a triangle, usually underneath a cylinder-shaped body. The advantage of this arrangement is that it allows the robot to turn on the spot, making it easy to control. Remember the degrees of freedom? A wheeled robot on the ground has just two: the ability to rotate, and the ability to translate, or move in the direction it is pointing (well, sometimes in the opposite direction to the one it is pointing in, but the idea is the same).

Some robots can be controlled directly in terms of rotation and translation—you can tell them to "rotate 45 degrees left" or "translate at maximum speed." Other robots have their wheels controlled directly. To rotate clockwise, you might turn the left wheel forward, at the same time as turning the right wheel backward. Rotating 45 degrees means working out how fast to turn the two wheels, and for how long, and is never going to be all that accurate. Then again, if I asked you to turn left exactly 45 degrees, you too might find it a little hard to estimate accurately.

As every driver knows, cars, unfortunately, do not rotate on the spot, useful though this might be when trying to park. The steering wheel does not quite give you control over the two degrees of freedom, so that turning around involves at least a three-point turn. This

LEFT

ROBOT EXPLORERS
In the aftermath of the attack on the World Trade Center in New York in September 2001, and the collapse of the two towers, the Center for Robot-Assisted Search and Rescue organized a number of different robots for exploration and assessment of the ruins.

SEP 15 2001
11:25:01 PM

ABOVE

MOBILITY IS EVERYTHING
Dr Who's Daleks were formidable enemies—until they encountered a flight of stairs.

BELOW

MARSOKHOD ROVER
This planetary surface vehicle was originally built in the former Soviet Union but has since been developed by NASA. Its six wheels on movable axles allow it to climb over rocks 1.5 times the height of its deeply-ridged, conically-shaped wheels.

makes four-wheeled robots harder to control if they have to maneuver in tight situations rather than just going straight ahead, the way a car does on the road.

TRACKING OFF

The problem of getting a wheeled vehicle to move over rough terrain was also tackled a long time before robots were mobile—by putting a moving track around the wheels, as in tanks and some planetary rovers. A tracked robot generally takes more energy to drive than a wheeled robot because more of it is in contact with the ground, and this has implications for power conservation in an autonomous device. There's also an extra degree of mechanical complexity to worry about—and if you'd sent a robot all the way to Mars you certainly wouldn't be happy if one of its tracks jammed. On the plus side, tracks do prevent the robot sinking into soft ground.

But let good engineers loose on redesigning wheels and you could be surprised by the result. The Marsokhod Mars rover, originally developed in the former Soviet Union, has six wheels, each individually driven, with a suspension system that allows each one to move relative to the others. The wheels are each shaped like a cut-off cone and deeply ridged, giving them some of the grip a track would have had. All this allows the rover to get over surprisingly large rocks—up to 1.5 times the largest diameter of its wheels—though not at any great speed.

Moving around fast presents a new set of problems for robots—for example, starting and stopping. A large lump of metal on wheels may have a substantial mass, and moving fast requires enough acceleration to get over all that inertia. Then again, once moving, a dead stop is out of the question, raising all kinds of issues we'll look at later when we consider how robots avoid obstacles in their paths.

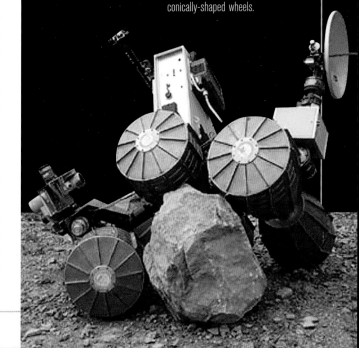

FEET FIRMLY ON THE GROUND

Wheels are a purely technological creation—they have never evolved on a living animal, probably because they require moving bearings to work. Legs are another thing entirely—they have evolved in a vast number of different species. So it was only a matter of time before roboticists started trying to put legs on their robots.

FOR A ROBOTICIST, LEGS HAVE BOTH ADVANTAGES AND disadvantages. Their big advantage is the mobility they give a robot on natural surfaces—those that are not artificially flat, like floors and roads. A leg only needs a small surface area to support it, and it can be lifted over smaller obstacles. Different legs can be put down on surfaces of different heights, and this allows the animal or robot to climb.

Legs are not as easy to control as wheels, however, precisely because they are more versatile. Each leg has at least one degree of freedom—and that's a leg that just swings from the hip in a single plane. If we give the leg a knee as well, that adds another degree of freedom; and if we allow the hip to rotate outward, that's one more. Ankle joints and toes start multiplying the degrees of freedom. More legs—more to control.

There are other complications with legs. For one thing, wheels stay on the ground all the time; but to walk, a leg has to lift off. As soon as it does this, it stops offering support or balance. Which leg should lift off, and when? How fast should a leg move, and how large a step should it take? If it has several degrees of freedom, the way the joints move will have a big effect—bend the knee more and the leg will stay off the ground for longer. In engineering language, it's not just the kinematics (where things go) that counts, but the dynamics (what forces at what times) too.

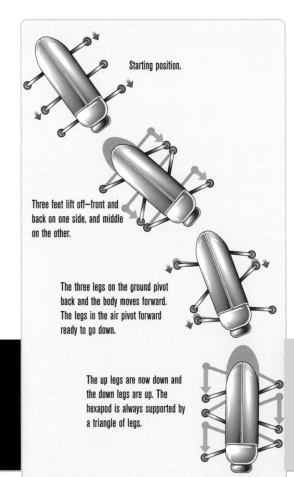

Starting position.

Three feet lift off—front and back on one side, and middle on the other.

The three legs on the ground pivot back and the body moves forward. The legs in the air pivot forward ready to go down.

The up legs are now down and the down legs are up. The hexapod is always supported by a triangle of legs.

ABOVE

SPEEDY BUG
Six legs work as far as insects are concerned—and cockroaches are the racehorses of the insect world.

RIGHT

TRIPOD GAIT
This six-legged robot employs the gait of the cockroach—even mid-step the insect is supported by a rock-solid tripod of legs still on the ground.

LEGGING IT

Looking at animals, some of these possibilities are bundled together into what we call a gait. For example, a horse can walk, trot, canter, or gallop: each of these gaits is defined by a characteristic set of leg orders and shapes, and presents different problems of balance. A galloping horse has all of its legs off the ground at one point during the gait.

So what is the optimal number of legs? In the natural world, mammals often have four, spiders eight, and millipedes many more, but six, as exemplified by hexapods—insects—is the most frequent number. Do six legs present a special advantage, or is this just accident?

Study of insects walking on smooth, open ground shows that they adopt a leg movement pattern called a tripod gait. In this, the front and rear legs on one side lift simultaneously with the middle leg on the opposite side. So the front and rear legs on the other side, and the middle leg opposite to them form a tripod, which gives a very stable balance. A robot with a tripod gait will not easily fall over, which is one reason why so many multilegged robots are hexapods. Insects have other gaits, though—in the wave gait, for example, the front pair of legs is moved, then the middle pair, and then the rear pair. Six legs have one other big advantage: if one or two get damaged, you can still get around.

THE COCKROACHES ARE COMING...

Impressed by the mobility of insects, researchers have made an effort to understand one in depth. The one observed most closely is the cockroach, which turns out to be the fastest thing on earth on six legs. Biologist Robert Full at Berkeley has been studying them for years, putting them on treadmills, measuring the electrical impulses in their muscles, and filming them on fast video. He has found that the legs move too fast for the cockroach to control them in the standard way—in fact, the muscles operate like a spring, so that the roach bounces along. Because it uses the tripod gait, it doesn't matter exactly where and for how long the legs go down; the tripod always keeps it stable. Researchers at Case Western Reserve University have gone for maximum biological accuracy, and their hexapods replicate as far as possible the ones they see on their roach treadmill. The day a robot roach bounces into Search and Rescue Operations may not be far off.

ABOVE

HERD IN FLIGHT
Different combinations of legs on and off the ground with force and shape gives a gait. A galloping mammal has all four feet off the ground at some point.

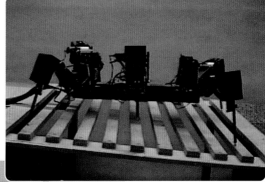

ABOVE

NIMBLE FEET
The second Case Western robot roach shows off its paces on a tricky surface.

RIGHT

HEXAPOD
The earliest Case Western robot cockroach looks simple, but it results from an intensive study of living cockroaches.

25

Six-legged robots may be stable, but it's hard to think of a robot cockroach as intelligent. Humanoid robots are the ones that grip our imaginations—and if a robot has to share our human environments, all designed for two-legged beings, there are some very practical reasons for having it walk around just like us. But it isn't easy...

TWO LEGS GOOD

THE MAJOR CHALLENGE FOR BIPED ROBOTS CAN BE SUMMED up in one word—balance. Even standing still on two legs requires some effort, as any human who has drunk too much quickly discovers! And walking means that, somehow, we have to balance on one leg most of the time. No wonder it takes children some time to learn how to do it.

The mechanics of walking have been intensively studied from all angles. Researchers in medical physics and orthopedics are even more concerned with the subject than are roboticists, since a thorough understanding of the processes is vital to helping humans who have problems with walking and to the design of artificial limbs. But in spite of all this research, we still don't know quite how we do it. More muscles seem to be involved than are strictly necessary—there are 246 solely in the lower limbs—and not just in the legs; the pelvis tilts during a step, moving the center of gravity downward, while the big heavy trunk has to keep itself above the moving legs. Even the smallest toe seems to play an important role in balance. Adding up all the degrees of freedom to be controlled in the human walk, we come out with hundreds—and remember we have to get the dynamics (the timing and the forces) right at the same time.

Of course, robots are usually made of metal and humans are not—and this means that just measuring the forces on a human will not produce answers for a robot, with different mass distributed in a different way.

ABOVE

FIRST STEPS
Considering the scale of the problem, a child learns to walk remarkably quickly.

STEPPING OUT

Despite decades of intensive study, the processes involved in the human walk are still not fully understood.

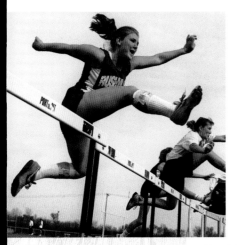

SWINGING ALONG

When an engineer looks at a biped, what comes to mind is a pendulum, a simple mechanism that has been studied for hundreds of years. When you take a step, one leg swings out from your hip, rather like a pendulum. The leg that stays on the ground swings as your body moves forward—but around the foot still on the ground, making it an upside-down or inverted pendulum.

One way of controlling robot legs is to use an idea known as the zero moment point, first conceived by Yugoslav researcher Miomir Vukobratovich, and since applied in many biped robots developed in Japan, including those developed by Honda. The zero moment point is where all the forces acting on the robot are in balance, though where this is depends on what it is doing, making it a sort of dynamic center of gravity. If this point can be kept inside the area—for example, the foot—supporting the robot, it will be in balance.

Honda's P2, and later P3, biped robots demonstrated the most natural and versatile walking ever, and could even climb stairs. This naturalness resulted from taking trajectory recordings of humans walking and climbing stairs—what is called *motion capture* in movies and computer games, where it is widely used. The clever bit is playing back the motions on a robot model, modulated to maintain its balance according to zero moment point control.

So, problem solved? Well, no. Other researchers need not give up just yet. The disadvantage of motion capture is that it does not explain why a given trajectory works. Nor does it explain how to select and blend together the various motions for new situations, such as walking across difficult terrain.

WALKING AWAY WITH YOURSELF

Just as with the cockroaches, it turns out that control is not everything when it comes to walking. A pendulum does not calculate where it should be next, it just swings. This idea has been used for many years to make little walking toys that set off down a slope without any power or special control being needed. Enter Passive Walking, first demonstrated by Tad McGeer, and being developed at Cornell University. Cornell has a 3D-walker that draws on a number of ideas. It has wide feet that guide its motion, soft heels that deal with the problem of determining just where the foot hits the ground, a counter-swing on the arms to reduce angular momentum effects around the walker's vertical axis, and arms that swing at appropriate times to reduce side-to-side rocking. This doesn't mean that brains and muscles are not needed to walk, but it shows that living things can take advantage of the way the world works and not do it all themselves.

SOME ACHIEVEMENT
The human lower limbs have 246 muscles—and we're still not sure how they all work together when we walk.

PASSIVE WALKING
This tinkertoy model from Cornell demonstrated that passive walking was more than just a good idea: put it on a surface and start it going and it carries on walking with no motors or batteries in sight— just the pendulum effect.

HOPPING, SWINGING, CLIMBING WALLS

Another look at the animal world shows that walking is not the only way to get around, even on dry land. One way of avoiding all those rocks is to jump over them. Another is to swing through the trees at a higher level. And then there are the vertical surfaces— how do we move along them? Geckos have one answer...

GIVEN THAT A ROBOT FINDS TWO LEGS a lot trickier to manage than six, moving to one leg might sound like asking for trouble. However, in the 1980s, the hopping robot, created by Marc Raibert at the MIT (Massachusetts Institute of Technology) Leg Laboratory, showed that this need not be so. Raibert's hopping robot, immortalized on video at the Boston Museum of the Computer, bounced like a pogo stick to maintain stability and could even turn somersaults. A combination of hydraulic actuators and pumped air (or mechanical springs) for supplying the bounce of the legs were used, both for this robot and for four-legged running robots.

The ability to hop and jump could be very useful indeed in some situations. Consider a robot that the US Marine Corps could use in a hostile urban environment, tossing it into an upper-storey window to carry out reconnaissance on the hop. NASA too is interested in an alternative to its crawling planetary exploration vehicles—a dozen hoppers could emerge from a lander to search in all directions. In the USA, Sandia National Laboratories has a hopper in a plastic shell the size of a grapefruit. Using a built-in compass and a gimbal mechanism with a movable weight, it can roll around to right itself after each jump. A small internal combustion engine with enough fuel for about

4,000 hops drives a piston into the ground, generating a leap three feet (1m) high and six feet (2m) forward.

Some animals can jump to impressive heights, and there are researchers modeling their robots on them—the robot cricket at Case Western and the JPL robot frog are just two examples. Neither of these resemble their animal inspirations in any respect other than their jumping ability, which is prodigious. Both also have wheels, enabling them to examine a small area in detail and jump over much larger ones.

ROBO-MONKEY

Toshio Fukada of Nagoya University, Japan has tackled an extremely challenging style of mobility—brachiation, or swinging from branch to branch. The Nagoya brachiator has 14 motors controling a fully articulated body, with colored balls marking key arm joints. These are used by a separate stereo-camera set-up connected to a computer that determines where the brachiator's arms are, updated 60 times per second. Using basic equations for swinging and knowledge of the distance between handholds, the computer sends the correct movement commands to the brachiator in real time.

Pendulum motion is important here, too. Initially, the robot kicks its legs for a pendulum swing to reach the first rung. If the brachiator fails to grasp it, an error is fed back, and it tries again. It shows that moving quickly can be easier than moving slowly if momentum is used.

STUCK ON YOU

Until now we have looked at broadly horizontal mobility. But what about vertical movement? Rubber suction cup feet and compressed air offer one way of doing this, but there must be others, or wall-climbing animals—the gecko, for instance—wouldn't be so successful.

Geckos use electrical forces—named van der Waals forces after the Dutch physicist who discovered them—to stick to a wall. This force results from the interactions of electrons in surface atoms and molecules. It is very weak for one atom, but can become quite large over an entire surface. Geckos' toes are covered in fine hairs—about 2 million on each toe. Structures at the tip of each hair roll onto a surface closely enough for van der Waals forces to work. A gecko's total electrical force could lift about 88 pounds (40kg), but by curling and uncurling their feet, geckos can peel their toes off a surface. Alan DiPietro, of iRobot in the United States, has created robot gecko feet, and found a way to drive his robot up the wall.

THE SLITHER FACTOR

Wheels, tracks, legs–how else can one move across the ground? What about snakes? Could we design a robot to slither like a snake?

ATTEMPTING TO DESIGN A ROBOT SNAKE MAY SEEM AN ODD enterprise—why would anyone want to? But as elsewhere, the motive of scientific curiosity (can it be done?) is augmented by the appeal of engineering practicality. Snake robots can operate in small, confined spaces, such as inside pipes, where wheels or legs would just get in the way; and they can wriggle through terrain that would present a problem for wheeled machines with more top-heavy profiles. Beyond this, a snake has a multipurpose shape that acts like a leg for movement, like an arm for grasping objects, and like a vertical mast from which it surveys the surroundings when it rears up.

Another attractive feature of snake-shaped robots is their modular construction; more links can be added as required to make a longer snake. If one of the identical segments breaks, it is far easier to install a replacement than would be the case for a conventional robot with many different kinds of components. The German MAKRO Project of 1997–2000 developed a multisegment robot to inspect the interior of sewerage pipes. The standard method of pulling a mobile through the pipe on a cable was expensive, and had limited success due to bends in the pipe and the distance between access points. A snake-shaped robot that could travel down the pipe autonomously was seen as a cheaper and much more effective way of carrying out these inspections.

Taking the MAKRO Project one stage further to produce a genuine "snakebot" would involve adopting serpentine motion, the form of locomotion used by snakes. As it turns out, there are more varieties of serpentine motion than you might realize. Grass snakes move by propagating a horizontal wave along their bodies, while the adder's wave moves vertically, perpendicular to the ground. Earthworms take a different approach altogether— they compress and then expand out along their length, in a technique known as extension. Cobras can do both extension and

23

horizontal waves. All of these methods work, but each has its pros and cons. Horizontal waves, for example, require little energy but produce a lot of friction with the ground. Vertical waves, on the other hand, cause little friction but take more energy to produce and, in terms of robot design, require vertical balance.

One of the earliest snakebots was made by Shigeo Hirose at the Tokyo Institute of Technology in the 1970s. Hirose worked on biomimetic robots long before the term had been invented. After making an intensive study of the movement of real snakes he produced a simple snakebot with serpentine motion by placing wheels under each modular section.

SNAKES IN SPACE

The whole field of snakebot development was given a great boost in the 1990s when NASA became interested in using snakebots for planetary exploration. The versatility and standard modular construction seemed ideal for space exploration, where keeping things light, flexible, and easy to repair are big advantages.

This work by NASA began with a *polybot*, developed by Mark Yim of Xerox Palo Alto Research Center in California, which had already produced modules that could be fixed together in variable lengths. Small off-the-shelf hobby motors were used in the snakebot's hinged segments, enabling it to move on a signal from the snake's main computer brain. Snakebot I was produced in a matter of days.

While Snakebot Mark I was directed externally, Snakebot II incorporated some autonomous behavior. This is no easy feat—remember the problem regarding degrees of freedom, which becomes more complicated the more segments you have. To deal with this, Snakebot II used two small microcontrollers in each hinged section to provide reflexes, taking care of simple but important jobs, and leaving the main snake computer to handle more important decisions.

In the future, NASA researchers hope to make snakebot muscles out of artificial plastic or rubber materials that bend when electricity is applied to them, to make something much lighter and more robust, like a car tire. Snakebots may soon be boldly going where no snakes have gone before.

ABOVE

SNAKE FACTOR
Researchers have found the snake a useful source of inspiration for locomotive ideas.

RIGHT

DRAINBUSTER
The MAKRO multisegment robot draws on the shape of a snake to get down sewerage pipes for inspection work.

FISH—WITH CHIPS

So far we've seen robots sally forth on land and in the air. But what about underwater? Are autonomous submarines the way forward, or can we learn a thing or two from sea-dwelling creatures?

CANNED TUNA
The coiled spring backbone of the MIT robot tuna—shown here with and without its Lycra skin—allows it to flex as it swims, just like a real fish.

BOX HEADING

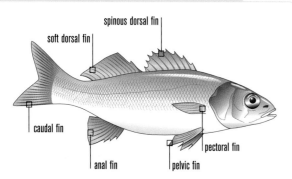

spinous dorsal fin

soft dorsal fin

caudal fin

anal fin

pelvic fin

pectoral fin

Each of a fish's fins has a precise aerodynamical function in achieving power, direction, and stability in the water.

IN THE 1930S, A RESEARCHER CALLED SIR JAMES GRAY investigated the swimming prowess of dolphins and tried to calculate how much muscle power was required for them to swim as fast as they do. Making a relatively small number of standard assumptions, he was surprised to find that ten times as much power was needed as seemed to be generated by dolphin muscles. This became known as Gray's Paradox and has stimulated much research into both dolphins and fish, in order to work out what allows them to perform so well. The pike, for example, accelerates at between 8 and 12 Gs, reaching 19 feet (6m) per second from a standing start. Either fish muscles perform much more efficiently than Gray thought, or alternatively something about the physiognomy of fish reduces the drag in the water that Gray assumed they had to swim against. So how about a robot fish as a means of investigating Gray's Paradox?

The first problem, of course, is that it has to swim underwater. This means that it must be tested in a tank, and also that a way must be found of keeping the electronics dry. A watertight body would be one approach, but the risk of leaks and the difficulty of getting at the components to change them makes this impractical, so the components themselves, as well as all the electrical connections, have to be waterproof. Then there's the way fish flex as they swim. Robot-fish researchers have usually chosen a spiral spring-like construction to achieve this effect, but there's the question of where to put the innards so that they don't get damaged as the body flexes.

Some of the best-known robot-fish have been constructed at MIT in the USA, starting with the robot tuna of 1995, built by designer David Barrett for his PhD thesis. Controlled by six servo motors each rated at 2 horsepower, it had force sensors at various locations along the path of its controlling tendons. A skin of Lycra minimized water resistance. This was not a completely autonomous swimmer, however: it was towed in the MIT tank to investigate swimming ability.

MIT later developed a robot pike. This had three bending segments—fewer than a real fish, but enough to make fishlike shapes in the water. It was made mostly of plastic in order to keep it light and water resistant, with the same sort of coiled-spring backbone as the robot tuna. Navigated by a human, it could demonstrate authentically fishy turns but swam more slowly than a real pike.

Mitsubishi Heavy Industries in Japan has spent four years developing robot fish with an accurate swimming action. They have developed a robotic sea bream, in which the tail fin and two pectoral fins are controlled by desktop computer. It uses an elastic fin, already developed for use in underwater vehicles, oscillating at between 0.2 and 1 cycle per second. This gives the robot a top speed of 0.82 feet (0.25m) per second. The battery that drives the fins is automatically recharged by a coil inside the fish, drawing power from an electromagnetic field that permeates the fish tank. Mitsubishi says its next step will be to build a robotic coelacanth, an ancient fish once thought to be extinct, and then other truly extinct fish from the pre-Cambrian era. An artificial aquarium will follow at a later date.

Researchers have used sea animals other than fish as models for their robots. At the US company iRobot there was interest in producing a legged robot that could scuttle along the bed of a river or lake—a robot crab. In this case, the application was a very practical one; they could be used for mine detection on beaches and underwater. Aerial, the robot crab, is completely autonomous and walks sideways on six legs. If one leg is damaged it can adapt its gait to walk using the remaining five. It is also invertible—if it gets turned over, it can walk the other way up.

Underwater robotics still presents a real challenge, but research into underwater creatures of all kinds provides continuous inspiration.

ON A WING AND A PRAYER

Sometimes crawling around on the ground is no good at all. Hummingbirds, dragonflies, and even grasshoppers manage to get up into the air—and micro-air robots are trying to follow them. Is it a bird, is it a plane, or is it a small robot spy?

HARD WORK
Bees use a novel method of generating enough lift to overcome their high body-weight to wing-size ratio.

IT ALL STARTED WITH THE US MILITARY. Imagine a soldier fighting his way through an enemy-occupied city. Wouldn't it be handy if he could send up a tiny plane—no bigger than a hand—to find out what was going on around the corner or inside a nearby building? Now imagine you're a TV producer covering a big sports event: How about letting loose a cloud of micro-air robots, each with their own on-board camera? Viewers on interactive TV could choose between hundreds of viewpoints.

The problem lies in getting such a thing off the ground and keeping it there. A standard plane relies on the difference in the airflow above and below its fixed wing to provide lift. This works well on an airliner with enormous wings, traveling at 500mph (800kph); but a tiny plane has tiny wings, and to be useful as a spy-plane it needs to move rather slowly. Even with a propeller, it's very difficult to get enough lift. Not to mention the problem that what constitutes a light breeze to a jumbo jet feels more like a gale to a tiny micro-air robot.

This brings us to the question of helicopters. They can achieve lift-off in very small sizes; the Pixel 2000 weighs only 1.7 ounces (48g) and has a single 11.7-inch (30-cm) rotor. But they are mechanically complex and very difficult to drive autonomously rather than by radio control, not to mention the problem of getting the rotor tangled in parts of the landscape. As Alexander van de Rostyne,

creator of the Pixel 2000 says, "I feel that the next step in shrinking down will need a new concept."

FLIGHT OF THE BUMBLEBEE

Conventional aerodynamics says a bumblebee should not be able to fly—it's too heavy given the amount of lift its wings generate. But this only holds for rigid wings. The bumblebee brings its wings together above its back so that they clap audibly, expelling air from between them. Air rushes in to fill the void as the wings are flung apart, and this generates immediate lift, as the air is already moving in the right direction.

At the University of California at Berkeley, the model for a micromechanical flying insect is the unglamorous blowfly, with a target size of less than 1 inch (2.5cm) from wing-tip to wing-tip. These will be biomimetic robots, which means that intensive study of the real thing—in this case, a tethered fly—will be used as a basis for the design. Piezoelectric actuators and flexible thorax structures can provide the necessary power density and wing stroke, and adequate power is supplied by solar cells.

Birds are the inspirational basis for some larger robots with flapping wings. Back in 1991, the University of Toronto produced the first ever Ornithopter, with 9-foot (3m) wings, powered by a conventional engine and controlled remotely. The experience has formed the basis

for a microversion, with a 6-inch (15cm) wing span, being constructed with the commercial technology transfer company, SRI (Space Robotic Initiative). The researchers hope that by studying insect flight and producing the new materials and energy-sources needed to make the theory work, micro-air robots will eventually be widely available for about $10. At that price, they could be launched in their tens of thousands.

SMART DUST

Make these robots even smaller, and they don't need to fly at all. Kristofer Pister of the Berkeley team is working on "smart dust"—a package of sensors, processors, and a wireless transmitter all designed into a grain of silicon so small that, like dust, it will hang in the air for hours once launched. Drifting on the wind, such grains could monitor the environment for light, sound, temperature, chemical composition, and a wide range of other information, and beam that data back to a base station miles away. So far Pister and his colleagues have managed to reduce them to not much more than the size of a pea, but with the rapid development of nanotechnology, millimeter-size robots are within reach. Many other tricky problems—power, communications, and programing—will have to be solved before smart dust becomes a reality, but who would bet against it happening before too long?

ABOVE LEFT

COPTER WITH CHICK
This helicopter, the Pixel 2000, built by Alexander van de Rostyne, is about as small as they go. To shrink any further, a different approach is needed.

ABOVE CENTER

EARLY BIRD
The first University of Toronto ornithopter flapped along on 9-foot (3m) long wings.

ABOVE RIGHT

BIG BIRD
Now Toronto is testing a full-scale, wing-flapping ornithopter, though with a pilot, rather than autonomously steered.

Lucy—A Robot Orangutan With Imagination?

NOT ALL ROBOTS ARE MADE IN UNIVERSITIES OR laboratories. "Lucy," built by Steve Grand, is very much a family robot. For Steve, the key question is how to build imagination into a robot brain.

After ducking out of teacher training college, Steve worked for the BBC and wrote educational software. He is probably best known as the creator and programer of the popular and wholly original Creatures computer entertainment. In this, the player starts with some eggs that hatch into virtual pets called Norns. The Norns have to be taught to behave sensibly and can also learn a small vocabulary. They breed, need to eat to survive, and can even catch strange diseases. More than a million people around the world "keep" Norns, and eggs and other paraphernalia are swapped over the Internet. Inside each Norn is a cut-down version of a virtual brain. Having succeeded with a computer-based creation, Steve has now set himself the challenge of making a real-world robot.

WHY A ROBOT ORANGUTAN?

"GOFAI (Good Old-Fashioned Artificial Intelligence) set out with the explicit aim of creating human-like intelligence, but got it all horribly wrong and wasted the best part of 50 years. As a reaction to this, a new kind of AI emerged; one that eschewed symbolic representations of thought processes and instead concentrated on more biologically-founded ideas. This new concept of Artifical Intelligence (and the loosely associated field of A-Life) focuses on understanding real nervous systems, especially those of insects and other invertebrates, and much simpler structures than the human mind.

But this backlash against symbolic AI leaves a huge gap between the biologically tractable, but not very intelligent, world of insects and sea slugs, and the overly-demanding, conscious, language-using world of humans. In this sphere lie most of the animals that inhabit our daily lives—dogs, cats, and birds. They are less sophisticated than humans,

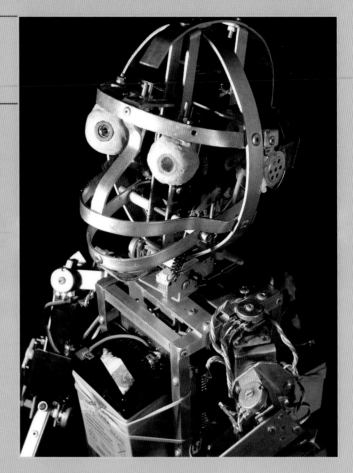

but their nervous systems are (in my view) profoundly different from those of ants. Such animals are highly intelligent, very adaptable, and (many would say) conscious of themselves and their environment. They're extremely complex compared to ants, but they don't use tools, form complex societies, and write research papers on AI, as humans do. It is their type and level of intelligence that interests us.

Maybe we should be building a dog—but Sony has been there and done that! On the other hand, for various reasons, early language is a useful thing for us to study and gives us important insights. Dogs can't speak, but primates like chimps and gorillas have been taught simple forms of communication, so an artificial primate seemed like a good

LUCY
Lucy doesn't have a lot of knowledge, but she does possess the ability to learn. Steve hopes that Lucy will learn like a real baby, and grow up in a family environment.

LUCY
Lucy's arms, neck, eyes, eyelids, and jaw all move, courtesy of servo motors configured like real muscles. In this first version, the "muscles" are not strong enough to support her weight, but later she will have more powerful ones so she will be able to learn to crawl. Seven separate processors control Lucy's senses, muscles, and communication between vision, hearing, and voice. Her brain is being developed on an external PC.

idea. We didn't want Lucy to be compared too closely with primate geniuses like Koko the gorilla or Panbanisha the bonobo, so we picked a slightly more distant relative—the orangutan. Plus orangutans look cute. Also, when we went to Toys-R-Us to look for soft toys that would give us anatomical inspiration, all they had was an orangutan."

ACTIVE PERCEPTION

The key to a robot brain lies in imagination: "Many people still think of the brain as a passive receptor of information. I think of perception as a much more active process. As conscious beings, we don't live in the real world—we live in a virtual world inside our heads. Most of the time, this internal world is closely synchronized to the external world—our model matches reality, tracks it, and predicts it. When we dream or when we imagine things (including making plans and rehearsing scenarios), we disconnect from the real world and let the model run on its own. Although the same mechanisms are at work in both cases, the synchronization with reality is missing when we dream or think. The model is the crucial thing: perception is an active process, in which we use this model to predict, hypothesize about, and correct the data fed in by our senses—filling in details when the data is incomplete and being surprised when reality fails to live up to the model.

A child's development is not about growing a better brain, but about growing a better model. The main difference between the ways in which a child and an adult perceive the world is that the child is less able to differentiate between fantasy and realism (not reality—none of us is aware of reality, only a realistic model of it), and hence the child is still capable of believing in, say, fairies. Lucy's brain is designed around a key set of hunches, first about how such a mechanism can be made using (simulated) neurons and biochemicals, and second, about how something similar might have evolved in nature."

HOBBYBOTS

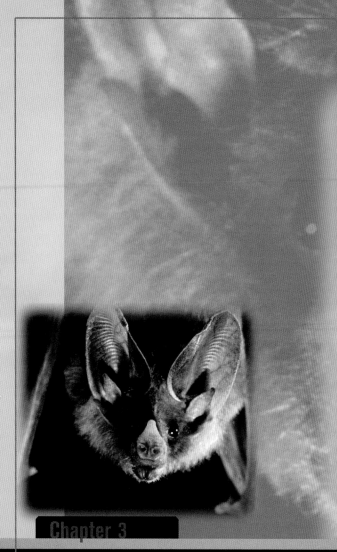

If mobility is what allows robots to do things, sensing is what allows them to do intelligent things. This chapter deals with the difference between active and passive sensors, and how they have been used since the time of the earliest autonomous robots in the late 1940s. We will come to understand why it is difficult for robots to see the way we do, and look at common ways for robots to perceive distance and coordinate perception and action in the basic problems of navigation and obstacle avoidance. Finally we will see how senses like hearing and smell are being investigated, and come to appreciate just how much these senses inform our view of the world around us.

If intelligence is all about interaction and doing the right thing in the real world, then sensors must be a requirement for intelligence. Without sensors, you couldn't tell you were in the real world, what you were doing, or the effect of your actions. For a living creature, moving around without any ability to sense changes in the environment would be a disaster waiting to happen—falling into a hole, or being eaten by something bigger would be but two of the nasty possible outcomes. And of course, the basic activities of feeding and mating also require sensors to tell you where the food or potential mate actually are.

Human sensing is dominated by sight, although we are all aware of the other four exterior senses—hearing, smell, touch, and taste. We are perhaps less aware of our interior senses—our sense of balance, for example, or our sense of body awareness, or proprioception, which is what informs us where our arms and legs are, and what they are doing.

One of the big debates in robotics since the mid-1980s has been about how a robot processes its sensory input. Before about 1985, it was generally assumed that the robot should build a model of its environment, which it could use when deciding what to do next. After all, psychological experiments have

Chapter 3

SENSING
THE WORLD

shown that humans appear to make such models—if I ask you to imagine your bedroom, then what you see with your mind's eye is an internal model. But many roboticists, led by Rodney Brooks at MIT, were unimpressed by the slow and easily confused robots of their day, and began to feel that model-building might not be such a good idea after all.

We share with other animals a rich sensory bond to our environment, with input and behavior coupled together very tightly in many cases—consider how your body tries to regain its balance when you trip over something, or how rapidly your foot hits the brake if you think you are going to collide with an obstacle when driving. Even quite lowly animals have a similar sensory richness, though the sensors are not always the same as ours. A cricket, for example, is covered with thousands of sensory hairs; each hair carries many sensory cells, some responding to mechanical input, some to chemical input.

So does the cricket build complex internal models? It doesn't seem likely. Brooks maintained that "the world was its own best model," by which he meant that linking sensing and behavior tightly together was more likely to lead to prompt and relevant action than spending a lot of time and effort building a model, which might be wrong or out of date by the time the animal had used it to decide what to do.

Robot sensory equipment, by comparison with the cricket's, is still often limited and impoverished: robots commonly have few external sensors and no internal ones. Yet by emulating the close coupling of sensors and behavior, even very simple sensing equipment can produce interesting and relatively complex behavior. This does not mean that roboticists have given up altogether on robots building internal models. But much more thought is going into when and how they are needed, and how they might be used in parallel with behaviors—those tight sensor-motor couplings that all animals need to survive.

OPPOSITE/BACKGROUND

NOT SO BLIND
Some animals have senses quite different from ours—this long-eared bat uses echolocation for the kinds of things for which we would use our eyes.

BELOW

EXTRA HELP
Human senses are good, but there are times when seeing in the dark is useful, and infrared night-sight glasses help to supplement our own eyesight.

RIGHT

BIG EYE
The 3,000 lenses in each of a fly's eyes can process changes in input up to 300 times per second, compared to 60 for humans. This gives them an acute sensitivity to motion, which combined with 360° vision is what makes swatting them so hard!

BEACON
The indoor scene under fluorescent lights looks quite different under infrared light—but the infrared emitter on the figure's buttonhole really stands out.

ACTIVE OR PASSIVE?

External senses can be divided into two types. Passive senses do not themselves affect the environment, but merely capture data provided by it. Sight and hearing both work like this. Active senses—like touch, for example—collect data by interacting with the environment. Other animals use active sensing much more than we do, which gives them a very different view of the world. What would it be like to be a bat? Robot sensing is still very primitive compared to living things, and much research is now going into making better sensors for robots.

PASSIVE SENSORS ARE WHAT HUMANS are familiar with—so much of our external sensory data arrives through sight or hearing. Equipping a robot with a digital video camera is one way of giving it eyes. How good these are at raw sight depends on the resolution of the camera—how many individual dots it can distinguish and capture in its field of vision; the frame-rate—how many pictures per second it takes; and the color discrimination—how many colors it can see.

Our eyes pick up full color in bright conditions, and because two eyes give us stereo vision, we can work out the depth of what we see. Other animals have eyes that put ours to shame. Where normal human vision is described as 20/20, a hawk's is equivalent to 20/5, meaning that the hawk can see from 20 feet (6m) what most people can see from 5 feet (1.5m). Falcons do even better, and can pick out a 4-inch (10cm) object from a distance of 1 mile (1.5km). Other animals see light outside our range—some fish can see infrared (IR), and penguins can see ultraviolet. But the really clever thing about animal vision is what they do with what they see, as we'll find out.

There are many more types of passive sensors around than you might expect. Many animals have sensory hairs— crabs use them to detect water currents, grasshoppers

use them to detect air movements, and blowflies have 3,000 of them on their feet, and use them to taste. Then there are animals that detect heat and electricity. Many snakes detect heat—a pit viper responds to temperature changes of two to three *thousandths* of a degree, while a rattlesnake can detect a mouse 16 inches (40cm) away if the mouse is 18˚F (10˚C) above ambient temperature. The duck-billed platypus has electric sensors in its bill that can detect 0.05 microvolts, and worker honeybees have a ring of iron oxide in their bodies that may be used to detect magnetic fields for navigation. Very few of these have been tried on robots.

In an active sensing system you learn about an object by bouncing energy off it. Echolocation—the capture and interpretation of reflected sounds, or echoes—is the animal world's best known example of such a system. Dolphins and bats are both echolocators: dolphins often swim in water with low visibility, while bats fly at night when vision is difficult.

Dolphins emit clicks at frequencies of up to 150KHz—about eight times beyond the normal range of human hearing—and convert the reflected sound into an acoustic image, allowing them to "see" in the depths of the ocean, or in murky estuaries. Experiments show that dolphins immediately recognize echolocated objects that they have previously seen with their eyes, or vice versa. Echolocation tells them the size, shape, direction, and speed of the object, and sometimes even what it is made of. Intriguingly, it also allows them to see inside some

LEFT

HAWKEYE
Although birds of prey, such as this eagle, have much smaller brains than humans, their eyesight is very much more acute—a falcon can see a mouse from a distance of 1 mile (1.5km) away.

BELOW

PENETRATING GAZE
Dolphins have good sight, but echolocation is more useful in deep or murky water. Like hospital ultrasound scanners, they can actually see inside other animals.

objects, the way ultrasound machines do. Dolphins do not need a machine to see the baby inside a pregnant mate.

For a robot, active sensing requires both an emitter to send out energy and a sensor to receive it. Dolphins don't seem to be deafened by their own echolocation clicks, but a robot has to be careful not to drown its sensors with its own emitted energy. Another problem is the way different surfaces react to an active sensor. Very smooth surfaces are poor reflectors of ultrasound, while very dark ones absorb, rather than reflect, infrared. As for corners, a common feature of indoor human environments, they are likely to disperse both infrared and ultrasound in several directions, hopelessly confusing their robot emitters.

The presence of noise—spurious unwanted energy in an electronic or communication system, often caused by interference—also plays a role. There are two main types of noise: white noise, the wide frequency spectrum disturbance that in a communication system gives rise to loudspeaker hiss; and impulse noise, caused by single momentary disturbances, audible as clicks. Noise in the system means that some of the received signal isn't real information, and sensor-processing robots must be able to distinguish between signals emitted by or reflected from objects and the background noise that threatens to drown them out.

KEEPING IT SIMPLE

Compared to dogs, dolphins, or even insects, the sensory equipment of most robots is really rather primitive. But oddly enough, even very simple sensors can produce surprisingly animal-like behavior in a robot.

ELSIE
Walter's first tortoise received its nickname because of its thick, protective plastic shell.

ANIMAL BEHAVIOR RESEARCHERS HAVE LONG BEEN INTERESTED in a type of behavior called taxis (pronounced tacks-iss), which is an animal's movement in reaction to some kind of stimulus. Depending on the stimulus, this can be described as phototaxis (movement toward light), pyrotaxis (to heat), or phonotaxis (to sound).

What must have been the very first robot to show taxis behavior was built in 1948 by William Grey Walter, a Kansas-born neurophysiologist who worked in Bristol, England. Walter built two three-wheeled robots, Elsie and Elmer, and though he described them as *Machina speculatrix*, because they explored their environment, they were familiarly known as tortoises, because of their protective hemispherical plastic shells.

Because he believed that the complexity of the animal brain arose from the degree of interconnection between neurons rather than their number, Walter hoped that it would be possible to build interesting electronic models using only a few neurons. The tortoises had just two, in the form of interlinked vacuum-tube amplifiers that drove relays controlling two motors—one to drive the front wheel and the other to control the steering. These artificial neurons were also linked to two sensors—a light-sensor on top of the steering column, which pointed where the wheel pointed, and a contact switch on the shell, which closed if the shell collided with anything, and made one of the amplifiers oscillate.

In scan mode, the tortoise would move in curves, the light-sensor turning with the wheel. On finding a light source, it would move toward the stimulus—exhibiting phototaxis—until at very close quarters a dazzle reflex took over, and the tortoise would back off and start scanning again. This produced a sort of light-dance. The light was mounted over a charging station, or hutch, and when the battery ran low, light-avoidance behavior would weaken until the tortoise would enter the hutch for recharging.

When a light source was mounted on each tortoise's shell, the robots would move toward and away from each other in a way that resembled some types of animal mating behavior. Walter later produced a new robot—he called it *Machina docilis*—which could be trained in much the same way as the Russian scientist Pavlov (whom Walter had met) had trained his dogs. This model added a connection between the two tortoise-sensors and a third sensor, which picked up the sound of a whistle. Walter trained the robot by blowing the whistle and at the same time kicking the shell to activate the contact sensor. Once trained, the robot would act as if it had hit an invisible obstacle every time the whistle was blown.

BRAITENBERG VEHICLES

Walter's work seems to have been largely unknown to roboticists in the United States, but similar ideas were raised much later by Valentino Braitenberg, in his highly influential 1984

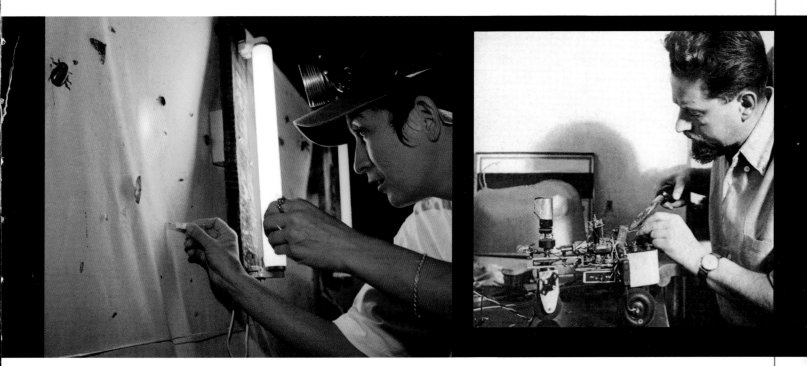

MIT Press book *Vehicles: Experiments in Synthetic Psychology*. He discussed a series of "thought experiments" of increasing complexity—though many of his vehicles have since been built, and Braitenberg himself has said that many of his experiments could be carried out with Lego's Mindstorms kit.

Early vehicles in the sequence are insectlike, with direct connections between sensors and wheel motors. Their behavior is created by changing the strength and pattern of these connections. For example, if a light sensor on one side of the vehicle is attached to a wheel on the opposite side, then as incoming light increases, the opposite wheel rotates faster and the robot turns toward the light. If it turns too far, the amount of light received drops, and the wheel slows down. Meanwhile, the sensor on the other side receives more light and drives its opposite wheel faster, so that the robot turns toward the light again. By wriggling from side to side in this way, the robot gets as close to the light as it can.

Braitenberg goes on to add multiple motors and sensors, cross-connecting their wiring and making some of them inhibitory. This produces extraordinary results—creatures that are still extremely simple, but show amazingly lifelike behaviors that can be interpreted as fear, aggression, love, and affection, combined with a wandering eye.

ABOVE LEFT

FATAL ATTRACTION
Phototaxis is exhibited by moths flying around a light bulb—sometimes with deadly consequences.

ABOVE RIGHT

ANIMAL MIMICS
Despite their relative simplicity, William Grey Walter's Tortoises were capable of producing quite convincing synthetic behavior.

HOW FAR DID YOU SAY?

A grasp of distance is essential in a three-dimensional world. Food at hand is better than food an hour away. A predator within striking distance is more of a threat than one on the other side of a field. A rock right in front of you is one you need to avoid. But how do you work out how far away things are?

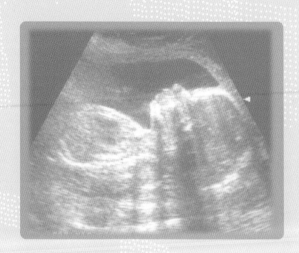

THE ANSWER LIES ON EITHER SIDE OF YOUR NOSE.

Each of your eyes captures one view, two slightly different images arrive in the brain, and an instantaneous computation later you know how far away the object is. The same thing can be achieved with two cameras on a robot. Work out the correspondence: for every point in one image, find the corresponding point in the other and then work out the difference between them. Given the differences between pairs of points, the focal distance of the two cameras, and their relative positions and orientations, some simple geometry will give the three-dimensional coordinates of all points—a process known as triangulation.

All this is perfectly simple in principle, but in practice it isn't quite so easy. There are many points in an image, and usually many images per second, making correspondence harder. What's more, both correspondence and triangulation computations assume an ideal pinhole camera, while in reality the low-cost imaging devices actually used are rather different. Cameras mounted on a robot are prone to being knocked out of position or moved by vibration over time, making an exact determination of their relative positions and orientations far from straightforward. To work effectively, cameras have to be calibrated before each use. Just imagine having to have your eyes adjusted each morning.

RANGEFINDING WITH ULTRASOUND

So robots often use much simpler methods to work out how far away things are. Sonar, or ultrasound, uses the echolocation idea: a transducer pings the object and waits for a return echo. The robot's controller times the flight—the period between the initial pulse and the return—and, knowing the speed of sound, calculates the distance from the object. Polaroid Sonar rangefinders, originally designed for autofocus cameras, have been widely used, and work from about 6 inches (15cm) to 35 feet (10.5m).

Here again, things are not as easy as you might assume. Sound pulses spread out as

they travel away from the robot, and may bounce off several different objects, while the type of surface a pulse encounters will have an effect on how much of it bounces back. Very smooth surfaces, such as office walls, or worse still, glass panes in doors, may not show up at all on sonar, unless the pulse hits them head-on. A glancing pulse may bounce off in the opposite direction, like an oblique shot in a game of pool. If it then hits another smooth surface it could eventually return to the robot from a completely different direction. On top of this, a robot needs to send pulses in several different directions to avoid hitting something just to the side of its intended path. With eight sonars in operation, it's a little harder to know which echo came from which pulse.

INFRARED ALTERNATIVE

For these reasons many robots employ infrared sensors rather than sonar. Infrared devices are small and they are cheap—any LED (light emitting diode) will do as an emitter. The fact that the beam moves at the speed of light makes accurate calculation of distance more of a problem, but has the advantage of lessening the effect on the calculation of the robot's own movement. While sonar is affected by smooth surfaces, infrared is affected by color. Black surfaces tend to absorb light (that's why they look black!) and sunlight contains so much infrared that it floods out infrared-sensing; so staying indoors is a good move, as long as you don't tangle with other sources of infrared, like desk lamps, or radiant fires.

Where real accuracy is needed, laser rangefinders are sometimes used. This uses the same basic idea, but the pulse is now a low-power laser beam. Because the beam remains tightly focused and doesn't spread, it is close to the ideal of a pulse that just bounces back, but lasers can also be used in more ambitious, and expensive, systems. In one set up, below right, a laser scan gives enough information about the shape of an object's surface that the technique is being used to make accurate replicas of 3-D artifacts, like statues for museums. The problem for a robot—apart from the cost—is that the beam moves slowly, which requires the robot to sit still for some time.

Whichever way we look at it, robot senses still lag way behind those of even the humblest animals. Or does a robot really need to know exactly how far away anything is?

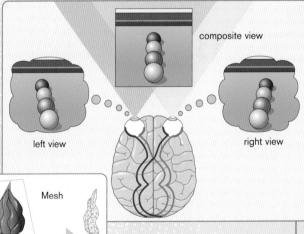

composite view

left view

right view

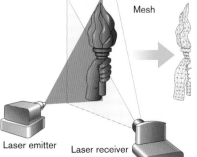

Mesh

Laser emitter   Laser receiver

SEEING IS BELIEVING

Humans are very visual animals, with binocular eyes and good color vision. A large part of the brain is given over to processing visual data, not to mention the processing that goes on first in the retina, in the back of the eye. The huge role sight plays in our lives is even reflected in the words we use: we say "I see" when we understand something. So what stops us giving robots the same wonderful visual ability we use so much?

WHEN YOU LOOK AROUND YOU, YOU ARE NOT CONSCIOUS of all the work your brain is doing. For a start, your eye is constructed so that the arriving image is upside-down. Your brain turns it right side up. An experiment once equipped subjects with "upside-down glasses" that inverted the incoming image. After a few hours, their brains adjusted and the world appeared right side up again—until, that is, they took the glasses off, when the brain had to adjust back to its original setting. All that wonderful reality out there is nothing more than the brain's interpretation of the stimulus of incoming light on the rods and cones in your retina—it's all happening in your head.

This effortless ability to perceive and interpret visual input makes it hard for us to understand the problems a robot might have. We look at the picture sent by a camera—the same picture the robot is getting—and we see the inside of an office with walls, a desk, and a chair. But what the robot sees is a rectangle full of thousands of little colored dots. Each of these dots (called pixels, short for **pic**ture **el**ement**s**) is one unit of information about the world, information that has to be interpreted before it means anything.

PIXELMANIA

The number of pixels in a single image is determined by the resolution of the camera. A low resolution digital still camera, for example, produces an image of 800 x 600 pixels, making 480,000 dots. Of course, a videocamera doesn't just provide a single image, but maybe 60 per second, making 28,800,000 dots per second in total. Depending on the number of colors represented, each dot can have up to 32 bits of information associated with it. And if we want our robot to have two eyes like us, we need to double all of that. That's a lot of data to process!

What does a poor robot do with all of those dots? Researchers in computer vision developed computer programs that process the information sequentially as follows: First

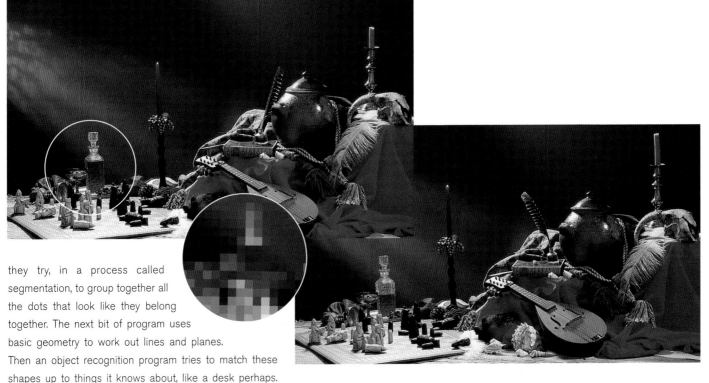

they try, in a process called segmentation, to group together all the dots that look like they belong together. The next bit of program uses basic geometry to work out lines and planes. Then an object recognition program tries to match these shapes up to things it knows about, like a desk perhaps.

This is hard enough work if the robot is stationary, but if it moves around, all the objects it encounters will look different from different angles, making it very difficult. While humans don't usually have any trouble recognizing a desk from two different ends of a room, no one knows quite how this is accomplished.

Humans can do other clever things from an early age. Even newborns will wriggle enthusiastically in response to anything that looks like a face. If you set up a screen in front of an older baby and move a toy so that it goes behind the screen, the baby's gaze will shift to the far edge, expecting the object to emerge again. Clearly we have quite a lot of built-in knowledge of how things move.

Some of the work that may help robots to see better tries to understand the way humans and other animals process visual data, rather than just throwing computer power at the problem. Maybe pattern recognition looks for particular shapes—as in faces. Focusing on what is interesting (things that move, for example) may help us to deal with the data overload. The pioneering work done by Carver Mead at Caltech (California Institute of Technology) on the silicon retina, continued by other researchers since, tries to produce chips that combine the input of visual information with processing the way we think the real retina does, rather than just throwing masses of data at a computer the way a camera does. So watch this space.

ABOVE, LEFT TO RIGHT

SPOT THE DIFFERENCE

You see a room full of objects. The robot sees a collection of dots. When the scene changes minutely, you have no trouble recognizing it as essentially the same picture. The poor robot has to begin all over again, checking out its groups of dots.

WHERE DID I PUT MY KEYS?

As we have seen, processing the huge flood of data from even one or two cameras on a robot seems a hugely challenging task. So how do we humans cope without all that processing power? Researchers have started looking much more closely at how we and other animals actually use our eyes, and how we avoid getting bogged down by too much incoming data. They call this active vision.

THERE IS A BIG DIFFERENCE BETWEEN the way animals (including ourselves) see and the way a camera records what's in front of it. For one thing, animals use their vision for vital tasks—finding food and avoiding threats from predators—so their visual systems have evolved to be sensitive to particular patterns. That's why newborn babies respond to a pattern that looks a bit like a face—locating mother is a survival issue. And note the word locating—animals actively seek out interesting and useful patterns, and don't just wait for them to turn up. Moving the head and the eyes to locate the object, and tracking it over time once found, all play a vital role in this information-gathering exercise.

The retina (the inside surface of the eye) has two kinds of light receptors, called rods and cones. Cones, present only in the middle of the retina, are color sensitive but have low resolution abilities. Rods have no color capability, but much higher resolution and much more sensitivity to movement. Their density is highest immediately around the cone area in the center of the eye, and drops off toward the edges. So if anything moves in the periphery of our vision, we quickly focus on it, moving our eyes to look

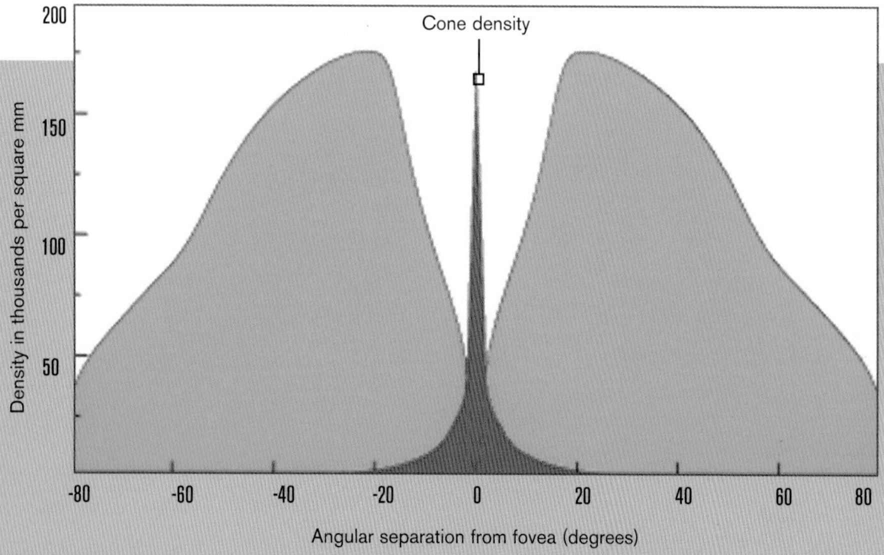

Cone density

Density in thousands per square mm

Angular separation from fovea (degrees)

LEFT

RODS AND CONES

The central part of the eye—the fovea—has all the cones and the highest concentration of rods. When we see something from the corner of our eye, we saccade by reflex, so that the fovea can take in the information. Cones only work in good lighting—that's why colors seem to fade at dusk—because better resolution from the rods is the most important thing under dim conditions.

at it with our highest resolution, and taking in its color characteristics. This is often called attentional focus. We only have to process all the extra color information contained in an object when we are looking straight at it.

EYES LEFT?

On top of this, our eyes are never still. Even when we look straight ahead, our eyes flicker. In fact, if we could keep our eyes perfectly still, the object in focus would fade out of sight, because the rods and cones need changing levels of stimulus to function.

More often, our eyes perform rapid, jerky movements called saccades. We make these movements several times per second when we are actively scanning, and they take place so fast that the brain has to cut out the visual input, or everything would look blurred. So the smooth visual experience we're used to is actually a set of jump cuts from picture to picture, with the brain editing out the gaps. Saccades are classified as voluntary, rather than involuntary movements, but it's thought they are produced with highly automated routines. We use them when performing

a visual search for a particular target, and also for ordered scanning, such as human text reading, like you are doing now. Finally, saccades are used to move the attentional focus when interesting new events occur in the visual field.

So how do we put an active vision system on a robot? The robot has to be able to interact with its environment by altering its viewpoint, rather than passively observing it, and by operating on sequences of images, rather than on a single frame. So we give its cameras their own degrees of freedom—generally using what is called a pan-tilt head, because it can rotate from side to side, and up and down, like a human head. Of course, if it sits on a mobile robot, the robot itself could also move. All this freedom of movement helps with some of the problems we looked at earlier. Tracking wins time for higher level processes (like object or task recognition) to complete; zooming in gives extra resolution for attentional focus, and zooming out gives context for task recognition.

BELOW

CLOSE WATCH
Study of the frog's visual system shows that it is specially tuned to objects about the size of a fly moving across its visual field. This is tightly integrated with the reflex that allows the frog to grab the fly with its tongue. The frog doesn't have to recognize that it has seen a fly at all.

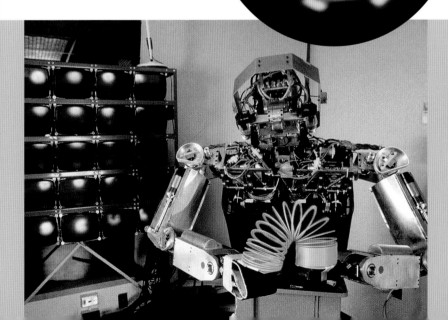

COG

Active vision systems seek to gather scene information dynamically and selectively, and one such system is used on Cog, a humanoid robot developed by the Artificial Intelligence Laboratory at MIT. The robot consists of a trunk, a head, and two arms and hands, and is equipped with sensory systems to mimic those of humans, including hearing, sight, touch, proprioception, and a vestibular system. This research group adopts the philosophy that the form of our bodies is essential to the development of thought and language, and that therefore humanlike artificial intelligence can only be created in a robot with humanlike form. Cog's sight comes from video cameras: to carry out saccades, it has to learn to map between its visual scene and its motor actions, giving it a basis for learning visually guided manipulation. Cog is a good example of the way that active vision makes vision just another robot behavior, rather than a separate sensing capability.

SMELLS GOOD…

Seeing and hearing play a big role in human lives, but we are much less aware of our sense of smell. How much do we know about how the sense of smell works? Could we put an artificial nose on a robot?

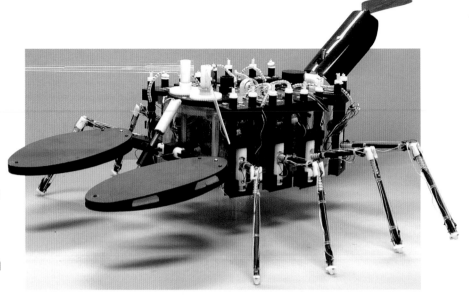

WHEN HUMANS AND OTHER MAMMALS USE THEIR SENSE OF smell, air passes through the nose and into two nasal cavities behind. These cavities are lined by a mucous membrane that contains many nerve endings, and these in turn are stimulated by different odors. The number of receptors is what determines how many different smells can be detected. It is estimated that humans have between 5 million and 15 million smell receptors in their nasal cavities; dogs, on the other hand, have between 125 million and 250 million.

This explains why dogs can detect miniscule levels of a substance in the atmosphere. It also highlights how they analyze smells in a way humans cannot. A trained dog cannot only search for a single substance but also for particular combinations of smells—police dogs are trained to recognize synthetic odors such as "burn victim" or "drowning victim." They can compensate for external conditions such as temperature, wind, and humidity.

We know that dogs recognize individual human scents, as well as individual dog and other species' scents. Dogs interpret other dogs' pheromones (chemical messenger scents) to gather important information regarding gender and receptivity to mating. They may also be able to detect mood and even illness, since it has been shown that changes in the body associated with both strong emotions and sickness may result in different odors being emitted—not so much a dog's eye as a dog's nose view of the world. It seems that humans react to pheromones too, but unconsciously. Recent experiments have suggested that humans may choose partners whose immune systems are different from their own based on the smell of their pheromones.

ABOVE

THE HUNTER
The lobster's sense of smell is so refined that it can sniff out a single amino acid that tags its favorite food, and locate its prey by following chemical plumes along the sea bottom.

TOP

ROBOT SNIFFER
Joseph Ayer's RoboLobster may not look very much like a real lobster, but the vertical sensing equipment on the top is designed to try out various theories of how real lobsters manage to follow chemical plumes so well.

A ROBOT NOSE?

We enable a robot to see by giving it a video camera, and to hear by giving it a microphone. But for a robot to have a sense of smell it needs an artificial nose, a type of sensor that is still under development. Though smell is one of the most primitive senses—and the only one directly connected to the emotional centers of the brain—exactly how it works in animals is much less well understood than sight or hearing.

Artificial noses are constructed from an array of sensing devices, each of which reacts differently to a particular compound, with supporting software that looks for characteristic patterns in the input in order to identify particular smells. At Caltech in the United States, researchers are experimenting with a 20-element array, using what they call polymeric sensors, made of polymer bases loaded with carbon black. The conductivity of the array drops where the smelled substance interacts with the polymer sensitive to it.

An array of millions of elements, like the animal nose, is still well beyond what can be engineered, so artificial noses are not nearly as versatile as real ones. Still, they can be designed with particular smells in mind, which is good if you are trying to detect explosives, gas, or rotten meat. But these devices are still too expensive and too undeveloped to be easily mounted on a robot, so roboticists have designed some sensors of their own.

Smelly, a robot under development by the University of Portsmouth in England, has two tubes containing a smell sensor sensitive to alcohol. Each tube has a small pump to suck in the odor, almost as an animal would sniff. The sensor is connected in a bridge circuit and its resistance changes when an organic compound is absorbed by the sensor film, allowing the concentration to be measured. Unlike an animal nose, the sensor has to be cleansed by passing clean air over it, and its temperature has to be controlled by a very sensitive circuit as the reading is temperature dependent.

FROM MOTHS TO LOBSTERS

Dogs are not the only animals that depend heavily on smell—so too do moths and lobsters. The silkworm moth can detect pheromones up to nearly 7 miles (11.5km) away in concentrations as low as 1 molecule of pheromone per 1017 molecules of air. A new European Union funded project that started in 2001 involves partners in the United Kingdom, Switzerland, Sweden, and France in an attempt to build a fleet of artificial chemosensing moths for environmental monitoring.

Lobsters smell their food using four antennae on their heads and sensing hairs on their bodies. The RoboLobster project that began in 1998 at MIT has tried to understand how they do this by implementing various strategies on a small-wheeled robot. Some other groups—for example at NorthEastern University in Massachusetts—are working on robot lobsters for mine detection.

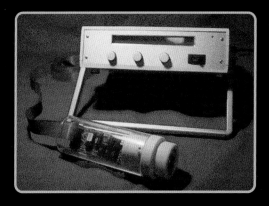

ABOVE

ARTIFICIAL NOSE
This artificial olfactory system has been developed at the University of Pisa in Italy, both as a handheld device for detecting olive oil aromas, and in the lab for environmental monitoring. Its sensing array is made of conducting polymer films that respond to chemical signals in gas form.

BELOW

CALL ME SMELLY
Smelly, from the University of Portsmouth in England, has smell sensors at the end of its two tubes, each equipped with a small pump to draw the odor in.

THINGS THAT DON'T GO BUMP...

If moving around under your own steam is what makes you a robot, being able to do it without falling over the furniture makes you an intelligent robot—or at least it shows you are aware of your environment and have enough sense to keep out of its way!

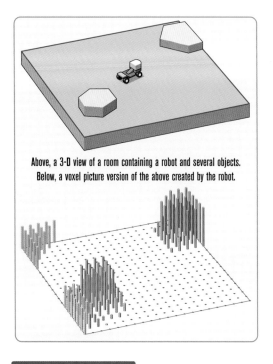

Above, a 3-D view of a room containing a robot and several objects. Below, a voxel picture version of the above created by the robot.

The robot's world is divided into voxels and sensor information used to decide which ones are empty and which ones are occupied. The voxel map shows that the robot has assessed the room above as follows:

Red—Probably occupied (and are indeed occupied)
Green—Probably occupied (but are not in fact occupied)
Blue—Probably not occupied
No bar—Certainly free space

THE ABILITY TO AVOID OBSTACLES IS A BASIC REQUIREMENT for any robot. Largely made of metal and often of human size, robots could do serious damage both to themselves and to their environment if they simply crashed into things. Industrial robots, which usually work "blindly" and without sensors, are carefully fenced off so humans who work in their vicinity are not at risk. Despite these safeguards, people are occasionally killed or injured when they venture inside the protective barrier, to carry out maintenance, for example. To be truly safe around humans, robots need sensors.

One way of dealing with obstacles is to compile an accurate map of your environment, and maintain an exact record of your location on it. The map will tell you where you can move safely. But how is such a map obtained? A robot that shares a human environment has to cope with change—desks and chairs may be moved in an office or meeting room, and people may also be moving around in the same space. Even if your map is accurate to begin with, it will quickly become useless unless it is updated with sensor information.

Fortunately, a robot navigating across an office doesn't have to know what the objects it is trying to avoid actually are, so identifying everything the way a human would simply isn't necessary. Let's imagine a robot operating in a room 20 feet (6m) by 10 feet (3m) by 8 feet (2.4m)—the size of the room isn't likely to change, so the robot could hold this information in its memory. The robot now divides this model into tiny cubes, say 3 inches (7.6cm) to a side. (These are sometimes called voxels, a 3-D version of the pixels, or 2-D picture elements, we encountered earlier.) It can also be assumed that the robot knows its own starting position in the room.

Next, the robot carries out some active sensing, using ultrasound, infrared, or laser. This gives a distance reading to the nearest object in the direction the sensor is pointing. All the voxels along the path taken by the sensor beam until it hits the object are marked as "empty." The voxel at the distance indicated by the sensor (that is, where the object is located) is marked as "occupied." As this process is repeated, the robot builds up what is called a volumetric map of the room, from which it can determine where it is safe to go. Modeling

the whole room would be cumbersome and time-consuming, but fortunately it is usually sufficient to focus on the area near where the robot is moving. The resultant map is often flattened into two dimensions, giving a horizontal slice through the room at the height of the sensors, and this is known as an occupancy grid.

THROWING THE MAP AWAY...

There is, of course, no evidence that an animal would avoid obstacles like this, and even a 2-D map takes time and effort to build and update. Moreover, determining your precise location on the map to start with is actually very difficult, as we will see later. Rather than making maps, the active sensing approach argues that sensing and behavioral response are linked, so that sensor readings can be used to control the robot's forward speed and rotation in different time frames. The robot could have a link between the two, which says "the closer the sensor reading says you are, the slower you go, and the more you turn away from the obstacle." The robot can do this without having any map at all, much as you would if you were driving a car around a dark parking lot with other cars parked all over the place.

This is a cheap and simple method of avoiding obstacles, and it can work very well. But because it is essentially a reflex, it can, on occasions, clash with other types of behavior, as in the example shown right.

IF ALL ELSE FAILS...

However a robot achieves obstacle avoidance, there's always a chance that something will go wrong and a collision will occur. This is one reason why most robots move around slowly, by human standards—if they do hit something, they must be able to sense the collision, and stop. To do this they employ contact sensors, trip switches activated when an obstacle is touched. These contact sensors may be attached to a bumper or, for more sensitivity, to flexible whiskers.

ABOVE

The robot wants to reach a door behind the desk. Its obstacle-avoidance reflex will make it slow down and turn away as it nears the desk. If it turns toward the wall, however, it is going to get stuck in the corner where the desk meets the wall. Without a map, it cannot know that if it turned the other way it could get around the desk.

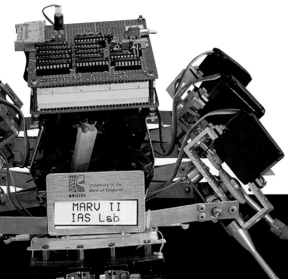

RIGHT

FEELING ITS WAY
Marv II, a hexapod robot developed at the University of the West of England, has flexible whiskers that inform it when it comes into physical contact with an object.

WHERE AM I?

We have seen that if a robot has a map, and knows where it is on that map, it can plan its journey to avoid any obstacles en route. The same system would also help the robot to navigate—plan a route to its intended destination. But determining exactly where you are on a map turns out to be quite tricky—roboticists call it the localization problem.

THE FIRST SOLUTION THAT OCCURRED to roboticists, and still one of the simplest, is called dead reckoning. The basic notion here is that if you know where you are to begin with, all you have to do is keep track of how far you have gone, and in which direction, and from a combination of these three elements you can calculate your position at any given time. Sailors have employed this method for centuries, using a magnetic compass to note their heading and a combination of sand or water clocks and knotted ropes to measure time and speed.

On a wheeled robot, the distance traveled can be measured by odometry—counting the turns of the wheels. Since you know the circumference of a wheel, knowing how many times it has turned will tell you how far you have traveled. The simplest odometer gives a pulse for each wheel turn, and something like a Hall Effect sensor, used in a computer's floppy disk drive, can be used to good effect. A more accurate odometer will have encoders like this all the way around a wheel—with eight, for example, a distance down to one eighth of the circumference can be measured simply by counting the pulses.

The direction the robot is facing can be worked out by putting an odometer on both left and right wheels and measuring the difference as it turns: to turn right, the right

wheel has to turn more than the left, and the difference should give the angle. A magnetic compass measures the heading directly.

NOT THAT EASY...

One of the phrases you hear most often in robotics is "but it's not that easy in practice." Even mariners were aware that dead reckoning was not very accurate, though in their case they had winds and currents to contend with. But if a robot is on the wrong sort of surface its wheels may slip, while if it has legs, the nature of the terrain may make it impossible to take steps of equal size. Calculating direction from odometer readings is just not that accurate, and a small initial error in angle can become a much bigger one over a long distance. Meanwhile, compasses are affected by the many electromagnetic fields on the robot itself (all those motors!) not to mention those in its environment. The errors in dead reckoning accumulate, and the further a robot goes, the worse they get.

Perhaps beacon-based navigation offers a better solution? Here the robot's environment contains beacons at known positions, and the robot can compute how far away it is from the beacons, or at what angles relative to its own direction they can be observed. All you need to determine your location are three such beacons and a little trigonometry—if you use the distances it's called trilateralization, and if you use the angles, triangulation. Robots using dead reckoning have only to see the beacons often enough to correct accumulating errors. This was one reason for establishing lighthouses for ships.

The systems generally employed by ships and planes use trilateralization, and calculate the distances from the time of flight of a radio signal, as we saw in active sensors. Unfortunately, there is a choice to be made between low frequency radio, which travels long distances but isn't all that accurate, and high frequencies, which are accurate but travel much shorter distances. For really short-range systems indoors, ultrasound or lasers are ideal. But all of these options have the disadvantage that they require the environment to be specially adapted to suit the robots, by setting up beacons in known positions for them.

GLOBAL POSITIONING SYSTEM

One of the best known, and increasingly used, beacon systems is actually supplied by orbiting satellites, and is known as a Global Positioning System (GPS). Some 24 GPS satellites, funded and controlled by the U.S. Department of Defense, orbit the Earth at an altitude of 11,000 nautical miles (20,000km), each taking 12 hours for an orbit. They are continuously monitored by ground stations located worldwide, and transmit signals that can be detected by anyone with a GPS receiver. An accurate clock on board the satellite puts a precise time stamp on the broadcast signals. The time difference between transmission of the signal from the satellite and reception of the signal on the ground, multiplied by the speed of light, enables the receiver to calculate its distance from the satellite. Signals from four satellites allow the receiver to work out latitude, longitude, and altitude to an accuracy of about 60 feet (18m), or half that amount if errors are corrected by a ground station at a known location.

On its own, GPS is not the answer for robots—the accuracy is too low, and high-frequency radio signals are difficult to pick up indoors. But as the receivers become ever more affordable, it seems certain that they will be a big help, especially when combined with other methods.

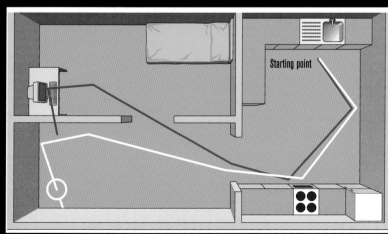

Starting point

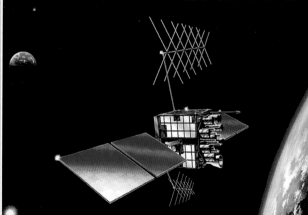

HOW DO I GET HOME?

When animals move around they mostly have a destination in mind: to the local waterhole for a drink, to a favorite hunting spot, or just home to the den or nest. If you know where you want to go, you need to plan a route, and then check that you are following it. So how do robots navigate?

WE KNOW THAT SOME ANIMALS PERFORM spectacular feats of navigation—homing pigeons, for example, can be released in a previously unknown location hundreds of miles from base and still get home. Birds on migration routinely cover thousands of miles, and salmon return across the ocean to the headwaters where they were spawned when their body clocks tell them it's time to mate.

Less well known are the impressive navigational abilities of insects such as bees and ants. Bees use an ingenious approach based on what is called optic flow. When you look out of the window of a moving vehicle, the landscape appears to be moving (which it is, relative to you). But not all of it moves at the same speed—closer objects rush past quickly while those in the distance pass by much more slowly.

This is how a bee judges distance, by the apparent speed of movement of an object's image, rather than by using complex stereo vision. Bees distinguish objects from backgrounds by sensing the apparent relative motion between object and background. When flying through a tunnel a bee stays in the middle by flying so that each wall seems to be moving at the same speed. When landing on a horizontal surface, it keeps the image velocity of the surface constant as it approaches, so

that it slows down. Foraging bees can work out how far they have flown by keeping track of how fast things were moving past them and for how long—a visually-driven "odometer" unaffected by wind, body weight, or energy expenditure.

Robots can use optic flow sensors to help them avoid collisions by keeping the optic flow of objects constant, just the way a bee lands on a surface. They can also use them for dead reckoning and, though still imperfect, this turns out to be much more accurate than wheel encoders.

PUTTING YOURSELF IN THE PICTURE

Researchers have looked at another possible bee or ant technique in their search for a better navigation system. Imagine you could remember what an object looked like from a certain position, almost as if you had photographed it on your retina at that point. Imagine now that you have the same object in view, though from a different position. By comparing the two images, you can change position so that the image looks exactly like you remembered it, and this gets you back to the position you were in when you first captured the image. Ships use a variation on this technique when docking, lining up dockside objects with the prow of the ship, and pilots landing planes manually also use it to line up accurately with the runway.

ABOVE

If you can navigate so that your remembered landmark corresponds to the one you can see, then you must be in the same position you were in when you remembered it.

Now let's suppose you have a set of objects, which are landmarks along your route. Using the technique above, you will be able to follow them back accurately, and so use them to navigate. You have produced a sort of map, albeit a relative one, and very different from the sort we were talking about before, where everything was represented with correct measurements and geometry. It's clear that the mental map we use to give directions resembles this much more than it does a printed map. Roboticists call these perceptual maps, because instead of being given to the robot by the designer, they are derived or learned by the robot from its working environment, much as they would be by an animal. What makes this idea hard to implement, though, is that because robot sensors are a good deal simpler than animal sensors, different landmarks often look much the same to a robot. This causes it to get confused about where it is on its perceptual map, and is known as perceptual aliasing.

REACH FOR THE SKY

What studies of pigeons and bees in particular have shown is that they do not rely on a single technique for navigation. It is known that pigeons can use the sun as a compass to work out where north is. On overcast days though, it seems they use an ability to sense the earth's magnetic field instead. Bees can also use the sun, but their backup mechanism is different. A small patch of clear blue sky is all they need. This is because light from the sky is partially polarized, and the plane of polarization in any part of the sky is determined by the location of the sun.

The lesson for robots is clear: intelligent movement in the environment requires good use of different sensors, and the resourcefulness to combine them—sensor fusion.

THE SUN COMPASS AND POLARIZED LIGHT

The neurobiologist Rudiger Wehner of the University of Zurich became famous for his work on the desert ants of Tunisia, who have to navigate in the heat of a fairly featureless desert. He showed that they measured direction by the sun and by the pattern of polarized light caused as sunlight hits air molecules.

One problem in using the sun as a compass is that it moves during the course of a day, so the pattern of polarized light changes as time passes. Using the sun and the polarized light it causes means keeping track of time—the desert ant relies on an internal circadian (daily) clock. It must also learn the path of the sun over the day—an ephemeris function.

A1: three-dimensional view of polarized light at 9am

B1: three-dimensional view of polarized light at 11am

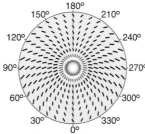

A2: the flow of polarized light around the sun at 9am (shown in 2-D)

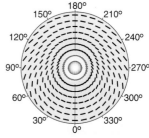

B2: the flow of polarized light around the sun at 11am (shown in 2-D)

Pigeons can find their home loft from up to 500 miles (800km) away. When released, they circle, looking for cues, and then pick a direction to fly home.

Uncontrollable Art

SCIENCE AND ENGINEERING ARE WHAT ROBOTS, on the whole, are all about: they are research vehicles designed to gain a better understanding of the mechanisms of living things, or to deliver technology that is robust and useful. But it is possible to see them in a totally different light, as does Simon Penny, Professor of Arts and Engineering at the University of California at Irvine.

Penny is an artist who uses technology as an integral part of his work. His current position, a joint appointment with the Claire Trevor School of the Arts and the Henry Samueli School of Engineering, follows an earlier tenure as professor of Art and Robotics at Carnegie-Mellon University. He is establishing an interdisciplinary graduate program (MA/PhD) in the Arts, Engineering, and Computation.

Penny has also written extensively on art and technology. He feels that the approach of the artist can compensate for what he sees as the "tunnel vision characteristic of certain types of scientific and technical practice." Where the scientist is concerned with the theory embodied in the robot and the engineer with how it functions, the artist's main concern is with human interaction, and the interface between robot and human. In producing his robot works, Penny thought first about how this interaction might work, and only then about the technology that would support it, where possible choosing simple, rather than complex, technical solutions.

Two of his early robot works are Petit Mal, designed in 1989 and built in 1993 with the help of several assistants, and Pride of Our Young Nation, built in 1990–1991.

PETIT MAL

Petit Mal was exhibited in many public spaces in the early 1990s. Penny says of it:

"The goal of Petit Mal is to produce a robotic artwork that is truly autonomous; that is nimble and has 'charm;' that senses and explores architectural space; that pursues and reacts to people; and that

PRIDE OF OUR YOUNG NATION

Pride of Our Young Nation is a meditation on machines of war and the psychology that motivates their construction. Its material construction echoes the archetype of a piece of field artillery, unchanged since the Napoleonic era, while its control system uses body heat detection and digital electronics. In this way it collapses the history of war machines into a single object.

The device senses the proximity of a visitor (using passive infrared detection of body heat). Process control circuitry initiates two actions. It tracks back and forth (locating the target), and "fires," its spiked conical barrel rotating and thrusting repeatedly and noisily as it does so.

gives the impression of intelligence and has behavior that is neither anthropomorphic nor zoomorphic, but that is unique to its physical and electronic nature.

Its public function is to present visitors with the embodiment of a machine 'intelligence' which is substantially itself, not an automaton or simulation of some biological system. More generally, Petit Mal seeks to raise as issues the social and cultural implications of 'Artificial Life.'"

The name of the work is significant, as Petit Mal describes a mild form of epilepsy, involving a momentary loss of consciousness.

"It is important that Petit Mal is just a little out of control—it is a reaction to oppressive theories of control so ubiquitous in computer science. It is an engineer's nightmare: although the mechanical structure is inherently stable, it has a chaotic motion generator at its heart—the double pendulum, an emblem of unpredictability.

The design process of Petit Mal sought not to design out unreliable behavior, but to capitalize on mechanical or electronic quirks, such as the dynamics of the double pendulum structure or the limitations of a sensor as generator of emergent behavior, or 'personality'.

Viewers necessarily interpret the behavior of the robot in terms of their own life experience. In order to understand it, they bring to it their experience of dogs, cats, babies, and other mobile interacting entities. In one case, an older woman was seen dancing tango steps with it. Another curious quality of Petit Mal is that, due to the desire of the user to interact, it trains that person to play; no tutorial, no user manual is necessary. People readily adopt a certain gait, a certain pace, in order to elicit responses from the robot. Also, unlike most computer-based machines, Petit Mal induces sociability amongst people. When groups interact with Petit Mal, the dynamics of the group are enlivened."

EXPLORING EMBODIMENT

More recently, Penny has been using intelligent vision systems to explore embodied interaction. In Fugitive (1995–1997), the movement of a human moving in a dark circular space is interpreted as an indicator of mood, and results in what feels like an instant projection of an image on a screen. In Traces, created during the late 1990s, real-time interaction between distant participants is created by real-time 3-D image (and sound) traces of their bodies being projected into each other's space.

⬆ PETIT MAL AND THE EIFFEL TOWER

Petit Mal's mechanical structure is a pair of nested pendulums of welded aluminum suspended from a pair of bicycle wheels on a common axis (a dicycle). This design is lightweight, economical (a full range of motion and control is achieved via only two motors), and self-stabilizing. The upper pendulum houses processor, sensors, and logic power supply. The lower pendulum houses motors and motor power supply. The inner pendulum functions to keep the sensors (more or less) vertical while the underframe swings due to acceleration. Batteries in both frames function both as power source and counterweight. It has three ultrasonic sensors paired with three pyroelectric (body heat) sensors in the front, and a fourth ultrasonic at the back. There is an accelerometer and two motors with optical encoders for motor feedback. These are coordinated by a single Motorola 68hc11 microprocessor.

Petit Mal will function autonomously in a public space for several hours before requiring battery replacement.

HOBBYBOTS

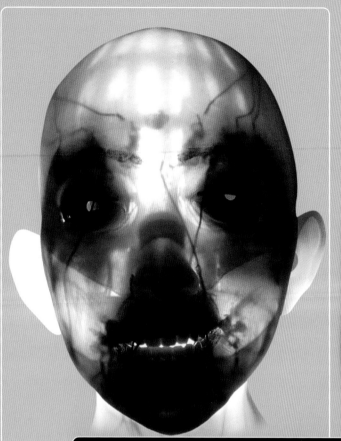

↑

THE FACE

This face robot was developed at the laboratory of Fumio Hara and Hiroshi Kobayashi at the Science University in Tokyo, Japan. The machinery beneath the robot's skin is used to change its contours, creating facial expressions. It does this by using electric actuators that change their shape when an electric current is passed through them. The actuators are made from shape memory alloys—metals which are easily deformed when the current is flowing, but quickly return to their original shape when it stops.

↗

WHO'S A PRETTY BOY?

MIT researcher Cynthia Breazeal holds a mirror up for her creation—the Kismet android robot. Kismet was developed for the study of action recognition and learning, particularly that of infants, and can display several moods through facial expressions, responding to visual stimuli like a baby. Here it is looking happy because it is receiving pleasing visual stimuli. When there are no stimuli, it shows a sad expression, and too much stimulation causes it to become distressed. Like a child, Kismet responds most favorably to bright colors and moderate movements.

While moving and sensing might be essential prerequisites for an intelligent robot, knowing where to move and what to do with what you sense is what really counts. As thinking animals, we might feel that intelligence has to be done our way, but a whole world of other creatures manages to get along fine without thinking the way we do. So what should we aim for in a robot?

Roboticists can learn a lot from very humble creatures. Rather than one deep-thinking robot, why not have lots of small, simple ones working together? Each robot would be much easier to construct; if one went wrong, the others would continue to function. This seems particularly sensible for robots designed to operate in a hazardous environment—in a collapsed building, say, or on the surface of Mars. Many heads are better than one, and sensors used over a wider area will locate hazards more effectively.

This approach has a further advantage: Bees, ants, cockroaches, and other small creatures have simpler brains than mammals, which makes it easier for us to understand how they work and to apply the principles to a robot. The creation of artificial neural networks (ANNs) has allowed us to program some of our knowledge about brains into robots, although we still have a very long way to go before we reach our goal. It's no exaggeration to say that a slug—because it can live autonomously in the real world—has more functionality than any robot yet designed.

Some researchers have taken a different track. Rather than building electronic brains, they are trying to connect an animal nervous system to a robot body. If we were able to do this successfully it would revolutionize the use of artificial limbs and enable the treatment of paralysis due to spinal injuries. The more sensationally-minded may see cyborgs as the next step, but given the versatility of living tissue, it is hard to see any advantage in this for anyone with a normally functioning body.

Chapter 4

PUTTING

THE LEARNING ROBOT

Animals' brains are difficult enough to understand, let alone program into a computer. So why not give a robot a powerful learning mechanism instead? Humans are not born with the ability to use their bodies properly—they have to learn how to walk. And babies only have a few innate behaviors—all the clever stuff like language and reasoning has to be learned. For this reason, many researchers are experimenting with learning mechanisms that allow robots to acquire skills and complexity the way we do.

The lure of a robot with human behavior and intelligence is still irresistible to most researchers. There are also practical reasons for trying to model human abilities. If robots are to function usefully in a human environment, there has to be some way of telling them what to do. A useful robot office transporter has to be able to do jobs on request. The ability to plan and reason seems indispensable for this kind of role.

So why not try to make a robot that reasons as well as a human? Does this take us back to the feared robot master race? It turns out that logic does have limitations in the real world of variability and change. Humans aren't always logical for good reasons, and the godlike logic we imagine for robots might not serve them at all well.

Integration is the name of the game—and the game is soccer. The RoboCup movement is using soccer as a testbed for research groups trying to integrate sensing, moving, deciding, learning, and cooperating. But it's proving a hard game to master! When a humanoid robot team can play a match against humans without disgracing itself, we'll know that robotics is really getting somewhere.

BACKGROUND

MONKEY BUSINESS
Chimpanzees display many intelligent behaviors similar to, though less developed than, our own. Should we aim at robots with human abilities, chimp abilities, or at something altogether simpler?

RIGHT

COORDINATED APPROACH
Soccer has turned out to be a more challenging test of robot intelligence than chess because it requires the body and the brain to work together seamlessly.

THINKING IT OVER

We have seen that many animals live successfully without doing much of what we might recognize as thinking. So does a robot need to think? What sort of activity requires thinking to succeed? Assuming we know what we mean by the word ...

WE'VE TAKEN "DOING THE RIGHT THING" as a mark of intelligent behavior, whether in animals or robots. It's clear that the amount of thought required for this depends on what sort of activity we're talking about.

Imagine that you are cooking and have a large cast-iron casserole on the stove. You grab the handles to pick it up, and you suddenly realize that it is very, very hot. Like it or not, your hands are just going to let go. This is an example of a reflex action—in the wake of the initial releasing stimulus the action has to follow. Reflexes are

very fast, and a good way of keeping safe, so it seems that robots should have some reflexes.

Reflexes are an example of innate behavior—they are hardwired into the nervous system and do not have to be learned. Not all innate behavior is as simple as a reflex, however—it can consist of a whole sequence of actions, as the following famous example shows.

Female spider wasps hunt for spiders, which they store in burrows to feed their young. A wasp digs a burrow, flies out and finds a spider, paralyzes it, drags it back to the mouth of the burrow, and then, before pulling the spider in, goes down the burrow to check that it's still empty. A researcher discovered that if you move the spider while the wasp is checking the burrow, the wasp will find its prey, drag it back to the burrow and go down again to see if the burrow is empty. Move the spider again, even a little, and the whole sequence will be repeated. One researcher did this 44 times before giving up—the poor wasp never did learn that the burrow was empty. Most insect behavior is thought to be innate like this.

GOOD REACTIONS

Not very much of human behavior is innate, but quite a lot of it is reactive—driven by what our senses tell us to do. Reaching to pick up a cup of coffee requires no more thought than feeling thirsty, but hand-eye coordination takes care of getting the cup and raising it to your mouth. Most of these motor skills work through reactions. When we are learning a new motor skill, we know we've nearly

REACHING OUT
Most human motor skills are carried out without conscious thought—hand-eye coordination carries us through.

RUSH HOUR
When we move in a crowd, we may think about where we're going, but the way we avoid other people is completely reactive.

BACKGROUND

SOCIAL ORGANIZATION
Most insects rely on innate behavior, but put a lot of them together and you get social intelligence and some surprisingly complex social behavior.

mastered it when we stop having to think about it.

Learning to drive a car, especially one with a manual transmission, is a good example of this process. At the start, it seems there are just too many things to be done at once—operating the throttle and the brake, changing gear, steering, and avoiding other cars. But as we get more fluent at these basic skills, much of this activity drops out of our consciousness. Our arms and legs seem to work automatically, while our concentration centers on planning our route or predicting what other drivers might do.

Reactions are an example of local decision-making—the action is determined entirely by the sensory input of the moment. Local decisions are both quick and always relevant to what is happening in the real world. So maybe robots just need reflexes and reactions?

The problems occur when what looks like the right thing to do at the time, later turns out to have been the wrong decison. And vice versa—what it later transpires was the right thing to do didn't seem appropriate at the time. For example, a robot floor cleaner shouldn't necessarily start cleaning a kitchen floor from the position it happens to be in when the task is set. If it does, it risks ending up stuck in a corner standing on the only dirty bit of floor left. What this suggests is that some problems can only be solved by taking a global perspective—and this is where thinking, in the human sense of the word, really starts to pay dividends.

IT ALL ENDS IN TIERS

It seems as though we'd really like robots to have three layers of actions. Right down at the bottom would be reflexes, which would be totally hardwired and very fast. Next would come reactions, where the robot would move according to what its sensors said: still pretty fast and not needing too much processor resource. Finally, and most slowly, would come global decision-making, involving a model of a whole problem—the strategy. There is a lot of agreement among roboticists that this sort of three-tier arrangement is a good idea and would work well on a robot. But the devil is in the details—and research is still trying to sort out exactly what makes up each tier, and how the tiers should work together.

BUILDING A BRAIN

If the brain is not like a computer, then just what is it like? Artificial neural networks are an attempt to model on a computer some of what we know about the brain—this is vital research for roboticists. Why not give a robot an electronic brain? Because as yet we only understand how very simple brains work ...

COMPUTERS TYPICALLY DO ONE THING AT A TIME—RETRIEVE A piece of data from memory, carry out some operation on it, put the result back in memory. We know that the brain doesn't work like that at all: it consists of very simple components called neurons, which although relatively slow are massively interconnected. Structures on the neurons called dendrites receive electrochemical stimulation from thousands—in some cases hundreds of thousands—of other connected neurons. If this input is greater than some threshold, then the neuron fires an impulse of its own. This travels out of the neuron on an axon and into terminators which stop at synapses. More electrochemical activity allows the impulse to jump across the synapse into the dendrites of thousands of other neurons, and so the process continues. Thus while individually relatively slow, the parallel nature of the reactions—thousands of neurons firing simultaneously and activating thousands of others—results in huge processing power. We still have much to learn about exactly what goes on in the different functional areas of the brain, but we do know that our ability to think, understand, and remember is distributed—there is no single neuron which holds the idea of "grandmother," for example.

Though the brain is not a computer, it is possible to model on a computer some of what we think the brain is doing. Artificial neural networks (ANNs) are based on a highly simplified electronic version of a neuron. The artificial neuron has inputs, and each input has an associated weight. All the active inputs are multiplied by their weights, and the result is summed. The artificial neuron has a threshold function, and if the summed inputs are greater than this threshold, the neuron fires and passes a value to all the other neurons to which it is connected, with this value multiplied by the weight in the connection. The value that emerges at the other side of the network can be thought of as the answer.

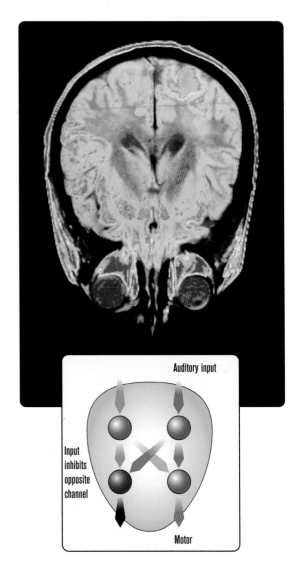

Auditory input

Input inhibits opposite channel

Motor

TOP

PARALLEL PROCESSOR
One thing we know about the brain is that it is massively interconnected.

ABOVE

BRAIN SWITCH
When a neuron receives an input signal higher than a certain value it fires a pulse to all the other neurons it is connected to.

TRAINING AN ANN

The interesting thing about artificial neural networks is that they are not programed but trained. One way of doing this is to provide some input data for which the desired output is known. The difference between what should be coming out and what actually emerges is used to alter the weights in the network; this process is repeated until the outputs are close to what is expected from the inputs. The power of ANNs lies in the fact that once the network has been trained on a range of inputs, it will respond correctly to inputs it hasn't seen before.

ANNs turn out to be good at recognizing patterns, even if those patterns are different from the ones they were trained on—they are used, among other things, to help recognize handwritten addresses on letters. This ability makes them useful in robots, where sensor data is often colored with the random variations caused by noise. Even small ANNs are able to hold the perceptual maps we considered in chapter 3 and use them for navigation.

THE ROBOT CRICKET

Researchers at the Universities of Stirling and Edinburgh in Scotland, and Lund in Denmark, have used a small ANN to control a robot they are using to understand how crickets find mates. Female crickets home in on males by listening to their chirping song. Sound reaches a cricket's eardrums—located on its forelegs—both directly and via internal tubes. The two sets of sound waves cancel or reinforce one another, depending on the location of the source, and this is how the female locates a male.

The robot cricket had four microphones feeding the digital signal into an ANN via a collection of amplifiers and delay lines. Two artificial neurons were constructed to have the same firing patterns as the cricket's; by adding only two more, the robot not only recognized the song but moved toward it. When the robot hears a sound from its right a signal passes down to its motor via the right-hand neurons, at the same time inhibiting the passage of any signal from the left-hand neurons, and the robot cricket moves toward the source. The ANN suggests that simple neural circuitry could be enough to control the behavior of a real cricket.

HEARING AID
The cricket has ears on its legs and special holes which allow the sound of a singing male to either cancel itself out or reinforce itself, depending on the location of the sound source.

NETWORKER
This ANN has simple processing elements, each with weighted inputs, a summing function, a transfer function, and links to other processing elements. The weights are set by training the network with inputs for which the right outputs are known.

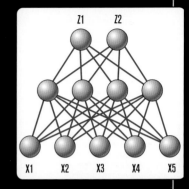

STIRLING CRICKET
This small robot may not look much like an insect, but the simple ANN that controls it allows it to behave just like a female cricket seeking a mate.

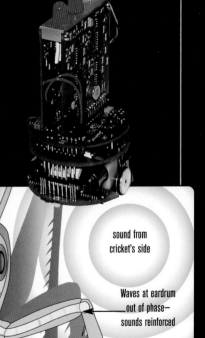

Holes

Waves in phase—sounds cancel

sound from cricket's side

Waves at eardrum out of phase— sounds reinforced

NEURO-SILICON?

Running artificial neural networks on a computer is only one way of trying to use the mechanisms of the brain. A different approach is to integrate real biological neurons with silicon. This could represent a path to replacing bits of the human body that have failed or been damaged with technology driven directly by the brain.

LIKE SATELLITES AND SPACE FLIGHT, both of which appeared in science fiction long before they were realized in fact, the integration of electronics and biology now being researched has already been foretold in the lurid imaginings of sci-fi writers. Sadly, the current state-of-the-art falls well short of the cyborg.

For years researchers have grown neurons—usually taken from rat embryos—in cultures, and have discovered that they spontaneously organize to form synapses. The problem has always been that, without a body attached, it is very hard to understand how they work in a living animal. What happens to our neurons when we acquire new motor

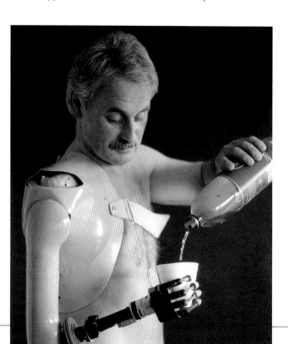

skills, for example, and learn to operate a car or ride a bike without having to think about it? Understanding what goes on in this "muscular memory" would allow us to construct ANNs for robots that might have something like our smooth physical skills.

The Pine Lab at Caltech has pioneered a method of siting cultured neurons on multi-electrode substrates—60 electrodes made of the transparent conductor indium-tin oxide (ITO) on a glass substrate—allowing their electrical activity to be monitored. A gas-permeable membrane made of Teflon protects the cultured neurons and allows them to be kept alive for two years or more. Steven Potter's group at Caltech have taken this further—they have connected the neurons to an animat, a simulated mouse moving around a virtual maze in a 3-D graphical environment. Electrical signals from the neurons are picked up by the electrodes and converted to movement commands for the animat, while its sensor data about where it is in its virtual world is fed back to the neurons. The next step is to understand the changes in the way the neurons organize themselves in response to these stimuli.

THE LAMPREY ROBOT

Sandro Mussa-Ivaldi and her group at Northwestern University Medical School in Chicago have connected biological systems to a real robot. They removed the brain and part of the spinal cord from a lamprey, an eel-like sea creature. Lamprey neurons are large and easily identified

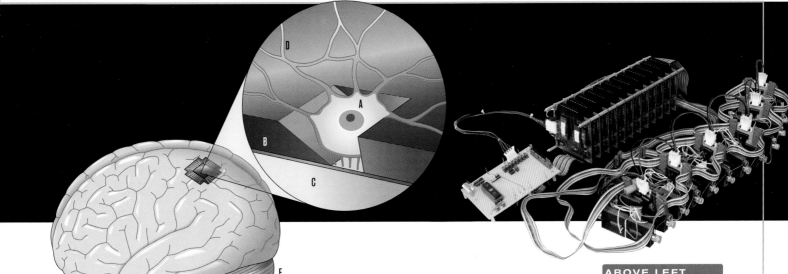

MONKEY SEE, MONKEY DO

and the tissue can be kept alive outside the animal's body for weeks in a refrigerated, oxygen-rich saline solution.

They made an electrical connection between a set of optical sensors mounted on a robot and the part of the lamprey's brain that allows it to distinguish up from down. The team argued that when the robot's sensors detected light, the lamprey's brain interpreted the incoming signals as a certain orientation in the water. The impulses from the brain that would normally have driven the eel's muscles were directed to the robot's wheels. Depending on where the electrodes were placed in the brain, the signals caused the robot to wheel toward or away from the light or to travel in a circle or a spiral.

Steve DeWeerth at the Georgia Institute of Technology in the United States is part of a group modeling the algorithms that move and control biological muscles using analog circuits. They are investigating how invertebrates, which have no brain at all, and simple vertebrates like the lamprey and the leech exercise motor control via their spinal cord. (The neural network that controls how the lamprey and leech swim is distributed in discrete ganglia that lie along the length of the animal.) The robots they build are helping them test out their theories.

A group at Duke University, North Carolina, embedded electrodes in areas of a monkey's brain known to be involved in motor activities. The electrodes recorded brain activity as the monkey learned to reach for small pieces of food on a tray, and the corresponding muscle movements were recorded. Data from many such actions was fed into an ANN running on a computer, and this learned the relationship between the brain signals and the muscle actions. When the monkey reached for the food, the ANN could predict its muscle movements and send instructions to a robotic arm, which would do just what the monkey was doing. The signals were even sent over the Internet to control another arm at MIT's "Touch Lab."

This idea is being developed at Carnegie Mellon University, where a team is creating an implantable neural interface that may eventually be used to control external devices and replace lost sensory input in humans. It is based on cultured neurons trapped, sustained, and controlled by microstructures as in the Caltech system: these neurons then act as a bridge to the host nervous system. The hope is that eventually, such implanted interfaces might be used to revolutionize the control of artificial limbs.

INSECTS, SPIDERS, AND CRUSTACEANS are almost entirely creatures of instinct. They have little ability to learn, and depend on built-in behavior patterns, acquired through evolution—a form of species-level learning. Fish, amphibians, reptiles, birds, and mammals, on the other hand, have individual learning capacities allowing them to modify their instinctive behavior.

So what type of process allows this kind of learning to happen? We have already seen the way that the weights in an ANN are changed through a training process. Remember that a set of training data is used for which the correct outputs are known, and the difference between the right answer and the ANN's answer is used to adjust the weights in the network. This approach is called supervised learning, and depends on a trainer. It would be nice if a robot could learn on its own, without supervision.

CHICKS

These youngsters instinctively crouch motionless when any moving object appears above them—even if it is only a falling leaf. Older birds have learned that leaves are harmless, and only react like this if something dangerous, like a hawk, appears.

REINFORCEMENT LEARNING

A different approach to learning could be summed up as follows: "Try something. If it works, learn it; if it doesn't, try something else." Rather than having a supervisor telling you whether what you're doing is right or wrong, you are rewarded or punished by your environment depending on whether your action is effective or not. Imagine a six-legged robot. By moving its legs in a given way, it either moves forward—the reward—or falls over—the punishment. The idea is to learn the movements that move it forward.

This basic approach can be worked out in more detail: you describe all the states the robot can be in, and all the actions it can carry out in each state. The robot then gets itself into each state and tries all the actions. The result of some actions may not be absolutely certain, but if the probability of an outcome can be worked out, in the end the robot will know which is the "best" action in any state. Better still, by giving each state a value, the robot can learn sequences of actions by summing up the rewards of all the states through which the sequence of actions takes it. The sequence with the highest combined reward would then be the "best". This approach avoids the problem of local decision-making, where the best action in the longer term may not look like the best action now.

EVOLUTION AND LEARNING

We know from selective breeding of animals that species change over time—the process of evolution over millions of years has produced our species, Homo sapiens, from an ancestor species common to us and the great apes. Our innate behaviors—smiling, the suck and grip reflexes of small babies—must also have evolved. So could we model the process of evolution and use this to produce robots with useful innate behaviors? Obviously we can't do this

LEARNING STRATEGIES

If it's so hard to program robots to behave intelligently, then why try? Nobody programs our intelligent behavior—we learn how to walk, to talk, and to interact socially. If a robot could have as powerful an ability to learn as we do, then couldn't it learn the same behaviors? Not only do individual animals learn, but species seem to develop over time. Researchers are trying to apply both types of learning mechanisms to robots.

for real, as robots don't reproduce, but it is quite possible to model on a computer what we think might be the main mechanisms of evolution, using what are called genetic algorithms, and then install the results in a real robot.

The genetic information that produces an animal, whether a cockroach or a human, is called its genotype, while the typical size, shape, and functioning of an animal is called its phenotype. Sexual reproduction takes place when two animals mate, and the two genotypes are mixed to give an offspring with a different genotype. It is thought that random mutations in the genotype also produce change, and if this turns out to give an advantage and help the animal better adapt to its environment, the mutation is likely to be passed on to the animal's offspring.

To model this on a computer, we must first create a population of genotypes, encoded as strings of numbers, each one slightly different. We have to decide on some measure of how good each genotype is, and pairs of genotypes are paired up with a probability depending on this measure: good genotypes get paired up more. Each pair of genotypes generates a descendant by copying some of its numbers into the descendant's genotype. We allow some copying errors and this acts like a mutation. As soon as we get a genotype that is good enough, we stop the process and use this genotype to generate a robot phenotype, which we can then build.

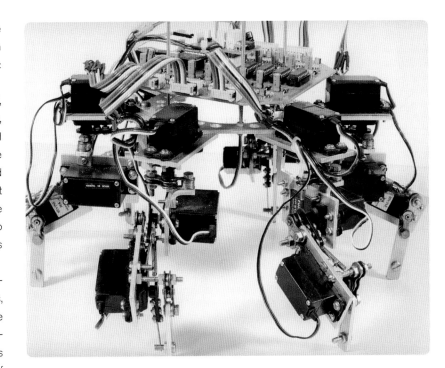

ABOVE RIGHT

TRIAL AND ERROR
Max, the six-legged robot from University of Aberystwyth in the U.K., has a lot of legs to organize. One way for him to do this is by trying something, and if it works, remembering it and using it again in the future.

RIGHT

REINFORCEMENT
The learning system inside the Agent records the angles (X) of Joint 1 and Joint 2 and how far along the ground the Agent manages to pull itself with these angles, which gives the reward, R. This information is then used to calculate how far to turn the two joints for the next move.

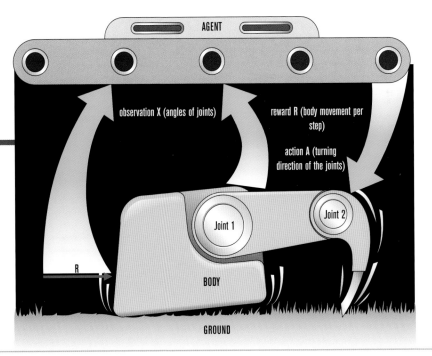

DO AS I DO

Humans and other mammals have developed and evolved by spending extra time rearing their young, who remain relatively helpless—but better able to learn new behaviors—for longer than the offspring of other species. A baby learns from watching its parents and trying to imitate what they do. As we saw with Lucy, bringing up a robot might be a way of having it learn our complex social behavior by imitation.

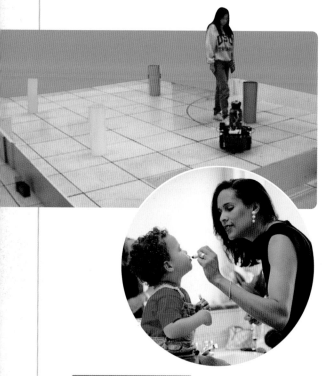

WE THINK OF IMITATION AS BEING KIDS' STUFF, AND NOT TERRIBLY intelligent. It turns out, however, that learning by imitation is a rare ability, found only in humans, and to a lesser extent in great apes, dolphins, and possibly some birds. It has become a topic of great interest to cognitive neuroscientists, animal ethnologists, and social scientists, as well as roboticists. For robots, learning by imitation comes somewhere between the supervised learning of an ANN and unsupervised reinforcement learning. Human "experts" can be used, making learning much faster. The experts just demonstrate how to carry out a task without explicit teaching, they can carry on uninterrupted as long as the "learners" can observe them.

If a robot is to learn by imitating, it first needs to observe what it is supposed to copy—some appropriate sensor input. This must be of high quality, or the robot will not have the necessary information to imitate successfully. To retain what it learns, the robot must remember this input. Next, the robot must have lower-level building blocks that allow it to recognize components of what it observes. It must already have some basic skills or behaviors to use as raw material for building sequences. These sequences are then used to build more elaborate ones.

Imagine learning to hop by watching another human do this. First you see the action. Next, you identify the fact that one leg is kept off the ground, while you "jump" on the other. Unless you recognize this, and are able to lift one foot and jump, you won't be able to imitate the hop.

This means that identifying a repertoire of building blocks (or primitive behaviors) is a key issue in robots' learning by imitation. Maja Mataric and her research group at the University of Southern California in the United States, are trying to classify natural human movements into a few primitive behaviors that can be represented with a set of parameters, and then use these primitives to classify and imitate observed movements and skills.

THE LANGUAGE GAME

We feel intuitively that an intelligent robot should be able to communicate in language. Could we have a robot learn language the way a child does? An experiment run by the Sony Computer Science Laboratory in Paris, France and the AI Lab of the Free University of Brussels, Belgium tried to do this in 1999 and again in 2000.

TOP

LEARNING BY EXAMPLE
The researcher moves between the colored targets in a particular order, and the observing robot, from the USC Robotics Lab, learns to imitate this sequence.

ABOVE

COPYCAT
Imitation is cleverer than you might think: a lot of what a child learns comes from imitating its parents.

The "robots" used were pairs of cameras with pan-tilt heads, located in front of a board covered with simple colored shapes in three or four different geographical locations. Hundreds of people took part in the two experiments, logging on to the website and using a software agent to briefly inhabit a camera and take part in what was called the "language game." Although there were only six or eight cameras, behind them was a set of hundreds of virtual robots all trying to learn to interact.

The language game was based on the sort of interaction you might have if you wanted to communicate with someone who did not speak your language. Using the camera you would point to an object, say the word for it, and providing the person could see the object, he or she could learn the word. Over time you would build up a common vocabulary.

For each language game, one camera robot was a speaker, and the other a listener. The speaker would focus on an area of the board and pick out a feature to talk about. If, for example, the board had a red square, a blue triangle, and a green circle, the speaker might say something like "red thing" if the red square was the topic. If there was also a red triangle, the speaker would have to be more precise and say "red square." In fact, the robots invented words by stringing together syllables, so that it might come out as "wabaku." The listener would try to guess the topic and point to what it thought the word meant. If this was correct, the listener would learn the new word.

If enough robots play this game for long enough, the experiments showed that they do build up a common vocabulary. But there's a catch: being robots running a vision program, what catches their camera eye on the board may be nothing like what a human would pick: "wabaku" may actually mean something like UPPER EXTREME-LEFT LOW-REDNESS. Like aliens, the robots may learn to talk to one another in a language humans do not understand.

BELOW

THE LANGUAGE GAME

Speaker
Perceive scene: speaker camera points at the whiteboard.
Choose topic: it focuses on an interesting item—say "yellow cube."
Conceptualize: it picks a feature to talk about—say "bright yellowness" or "four corners."
Verbalize: it makes up an arbitrary word for this: "wabaku."

Hearer
Interpret utterance: hearer checks if it knows this word and what it means.
Perceive scene: hearer camera points at the whiteboard.
Apply meaning: hearer locates the feature that corresponds to what it thinks the word means.
Point to referent: hearer focuses on the feature that corresponds to its meaning for the word, transmitting this to the speaker camera.

If speaker and hearer agree, then the word is part of their common language. If the hearer gets it wrong, the speaker points at the feature and the hearer tries to learn the new word.

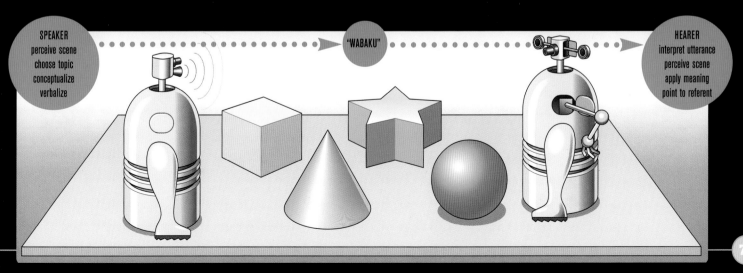

SPEAKER
perceive scene
choose topic
conceptualize
verbalize

"WABAKU"

HEARER
interpret utterance
perceive scene
apply meaning
point to referent

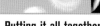

TOGETHERNESS

Though innate behavior is inflexible compared to learned behavior, social insects show that cooperation and coordination can act as substitutes for individual flexibility. From termite hills to beehives and wasp nests, we see complex constructions that have not been designed in the human sense but have emerged from the interaction of many termites, bees, or wasps. So, while termites have a limited range of fairly inflexible behaviors, interaction between large numbers of termites and their environment can produce a much higher than expected level of social intelligence.

BELOW

FRED & GINGER

The University of Salford's Fred and Ginger carry a piece of pipework together. Notice the square plates on top of each robot—these can move forward and back as well as left to right. Trying to keep them centered is what allows the robots to cooperate successfully in their carrying task.

WORK IN COLLECTIVE ROBOTICS TRIES TO draw on this approach—one advantage being that a small repertoire of innate behaviors is easy to implement on a robot and does not require a lot of computing power. So how do ants or termites manage to cooperate? We know they don't sit down and have a meeting—they just don't have the brain for it. Research has found that many of these apparently complex outcomes require no separate communication at all—if one ant changes the environment as it goes, the environment itself gives a message to other ants. A special term—stigmergy—has been coined for this communication through environment.

When an ant finds food, it leaves a chemical trail of pheromones on its way back to the nest. Should other ants who happen to be wandering around looking for food find such a trail, they tend to follow it. They too find the food, and deposit more pheromones on the way back to the nest, making the original trail stronger. The stronger the trail, the greater the number of ants that find it, and the more pheromones that are added to it. Eventually the food runs out, no more ants leave a pheromone trail, and the scent dissipates.

STIGMERGY AND THE QUICK WAY ROUND

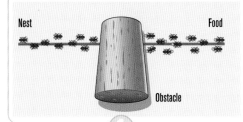

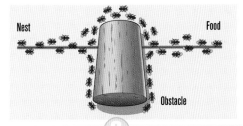

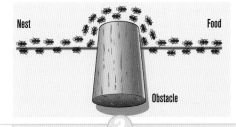

1 When an obstacle blocks the ants' path, they have to go around, some one way, some another, laying pheromones as they go.

2 Because the ants get around the shorter path more quickly, more ants follow the route per second, and a stronger pheromone trail is laid.

3 This gradually attracts all the ants to the quickest route.

Various groups, such as one at the University of Lausanne in Switzerland, have demonstrated stigmergy using a number of identical robots in a small area scattered with pucks. The robots have behaviors—driven by their sensors—to move straight ahead, turning at walls, until they end up pushing a puck. When they get close to a second puck, a "back-up-and-turn-away" behavior takes over, leaving the pucks in a cluster. Without any communication, the robots end up collecting all the pucks into one pile. This happens because when a robot approaches an existing pile directly, it adds the puck it was already pushing to the pile and turns away. Occasionally, a robot approaches an existing pile at an oblique angle and takes a puck away, but over time the desired result is accomplished. As with the ants, the stigmergy allows the robots to affect each other's behavior.

WORKING TOGETHER

Ants collecting bits of food individually may not seem all that much like cooperation. But stigmergy can also be used for robots working together more closely. In the 1990s, two robots at the University of Salford, England—Fred and Ginger, after the dancers—were able to carry objects around together without any explicit communication at all.

Each robot was fitted with a flat spring-loaded plate with a small robot arm attached to it. The robots' behavior was designed to keep the plate centrally located on top of their "heads." When Fred and Ginger carried an object together, they would sometimes move at slightly different speeds in slightly different directions. This would pull the plates off-center, which in turn would make them alter their speed and direction so as to stay in step with each other.

Imagine you are carrying a table with someone else. If they move too quickly, the table will pull away from you, and you will speed up to compensate. If they move too slowly, the table will dig into you, and you will slow down. This is how Fred and Ginger did their own furniture removal act without needing to tell each other anything at all. This is also a form of stigmergy—each robot forms a part of the environment of the other. Fred and Ginger did not need to know that they were carrying an object or that another robot was helping them—the movement of the table was enough for them to be able to cooperate.

SWARMS, FLOCKS, AND FORMATIONS

Many animals seem to move in an organized way when they congregate in numbers. Whether we call these swarms, schools, flocks, or herds, it has occurred to roboticists that if we could understand how they work, robots could swarm, school, flock, and herd, too.

ANIMAL FORMATIONS ARE FURTHER CASES OF COLLABORATION with very little communication. When geese fly in an arrow-shaped formation, it's not because they've had a vote and chosen a goose to lead the arrow. The shape is a good one for long-distance flights because the geese on the outside of the formation reduce the air resistance for those flying behind them. This also means that the goose at the tip of the arrow is working harder than anyone else, and observation of geese in flight has uncovered the fact that the lead goose changes—as one gets tired, it falls back into the formation to be replaced by another. It seems that small variations in the natural flight speed of each goose is enough to produce the arrow, with extra resistance at the edges slowing down stronger geese and making it easier on the inside for weaker ones.

So what behavior for each individual animal allows them to produce some of the complex formations we observe? Craig Reynolds, a creative programer now with Sony Computer Entertainment, developed an impressive model of flocking in 1986. Reynolds programed his animated creatures, which he whimsically called "boids," to follow three rules. The first separation rule was not to get too close to anything, including other boids—the kind of obstacle-avoidance we have already seen with robots. The second alignment rule was for each boid to try to match its velocity with that of boids around it. The third cohesion rule was always to move toward the average position of local flock-mates. Reynolds has animations (see www.red3d.com/cwr/boids/) which show how scattered boids form a flock and flow around obstacles in their path.

One interesting application of Reynolds' work has already made it into the movies: Disney's *The Lion King* and Steven Spielberg's *Jurassic Park* featured stampedes of animated creatures whose autonomous paths were produced using Reynolds' boid rules.

ROBOTS IN FORMATION

The same ideas can, of course, be applied to robots. One advantage of flocking is that robots can cover a wide area while remaining in sensory contact with each other. This is just what is

needed for search and rescue. The Sandia National Laboratories in Albuquerque, New Mexico, have been developing swarming robots to look for skiers lost in avalanches, though in fact the original motivation was finding the point source of a chemical or biological attack. Each robot continually informs others of its position and the strength of the signal it is receiving from the source—the buried skier, for example. The streams of information allow each member to continually refine its search, and Sandia's simulations showed this located the source four times faster than any other published technique.

Formation flying is also being taken way beyond the goose level by Space engineers. In the summer of 2001, NASA launched the Earth-Observation 1 mission to test out formation flying by satellites. These were effectively robots, reacting to each other and maintaining a close proximity without human intervention. They could autonomously react to each other's orbit changes more quickly and efficiently than would have been possible by traditional ground control. Scientists might obtain unique measurements by combining data from several satellites rather than flying all their instruments on a single expensive satellite. This enabled the collection of different types of information—such as stereo views or simultaneous data of a ground scene from different angles—that a single satellite could not provide. If one satellite had a problem, the others could reconfigure to work around it.

The European Space Agency is working on an even more ambitious project called Darwin, with which they hope to look for earthlike planets over vast distances. A technique called interferometry allows a number of smaller telescopes to combine their individual signals to mimic a much larger telescope. The idea is to launch six robot space telescopes that will fly in the tight formation required to make this happen, resulting in the most sensitive telescope ever known. Boids of a feather?

SEARCH PARTY
Sandia Laboratories in Albuquerque, New Mexico have been investigating the use of small swarming robots for search and rescue operations.

TEAM SPORTS

Swarms may have their uses, but humans normally work together in teams. The difference is that, in a team, not every member carries out the same function, but usually has a specific role related to his or her particular capabilities. Could we produce robot teams? As a way of understanding the problems and trying to arrive at solutions, research groups throughout the world have come together in the RoboCup competition—in which teams of robots compete in five (or so)-a-side soccer matches.

IN 1997, HIROAKI KITANO, THE PROMINENT Japanese roboticist, established RoboCup (www.robocup.org) as an international research and education effort. He wanted to stimulate research in artificial intelligence (AI) and robotics by providing a standard platform on which a wide range of technologies could be integrated and assessed. The ultimate challenge is to develop, by 2050, a team of fully autonomous humanoid robots that can win against the (human) world soccer champions. This parallels earlier efforts to develop a chess-playing system that culminated in Deep Blue's success in the 1990s. Soccer, however, is a much tougher problem and it is not yet clear whether the aim will be achievable.

The RoboCup competition is split into a number of different leagues for different types of robots, as well as one played entirely as computer simulation. Tournaments have been held every year since 1997, with rules which are gradually evolving as research teams get better at their task. For example, in the 2001 Small League, wheeled robots no bigger than 7 inches (18cm) in diameter and 8 inches (20cm) in height competed five-a-side on a table-tennis-sized field using an orange golf ball. The game was played over two ten-minute halves, with a ten-minute break in between.

A fairly basic problem for robots—as will be remembered from Chapter 3—is finding the ball in the first place. Knowing their own location on the field presents a further hurdle. Then there's knowing where the goals are, which goal they are defending, and which they are attacking. Not to mention the problem of recognizing teammates (as distinct from robots on the opposing team). All in all it's a pretty tall order, and for this reason overhead cameras are still often used to guide the teams in the Small League. The Middle League involves larger robots and no overhead cameras, making it a rather slow and fumbling affair—kicking the ball is tricky for a robot with no legs, and headers are out of the question!

RoboCup events attract reporters and photographers (as well as up to 5,000 spectators), but one thing you will not see, is flash photography during a match. Not only does the flash

LEFT

SCRUM

Team strategy is still very primitive, and there is a tendency for all the Aibos in a game to head for the ball at once, just like small children playing soccer. In this way the game can often degenerate into a struggling heap of robot dogs.

CRUNCH
Two Aibos move into a fierce
tackle...

ABOVE

KNOW YOUR ENEMY
It is vitally important in any game
to be able to recognize your opponent!

LEFT

SPRAWLING
...which leaves one flat on its back.

on a camera disturb the vision systems used by robots, but the infrared focusing devices used by many cameras can act like beacons and fatally disrupt the robots' navigation systems.

LEGGED ROBOTS

Those involved in RoboCup saw Sony's introduction of the Aibo robot dog as a brilliant opportunity to establish a competition for teams of four-legged robots on a professionally engineered basis, and three-a-side Aibo teams have competed since 1998. The Aibo includes a vision system that is sensitive to pink, and so Aibo RoboCup teams play with a pink ball. By installing checkered patterns of pink and yellow near the top and bottom of each side of the field, the Aibo players manage to get some information about where they are.

Kicking a ball still presents a few problems—but then how would a real dog kick a ball?—and the best current technique

seems to be for the Aibo to go down on its front elbows and whack the ball with both front paws simultaneously. Nobody has yet been able to get a robot to "trap" the ball, however, so there is a danger that a strong "kick" like this will produce ricochets off other robots or the walls around the field.

BUT SERIOUSLY...

All this may seem quite funny (certainly watching Aibos play soccer can be a hilarious experience) and an odd enterprise for professional researchers. On the serious side, RoboCup has just started a Search and Rescue group to think up standard tasks in less frivolous fields, where the teamwork principles learned in robot soccer can be reapplied.

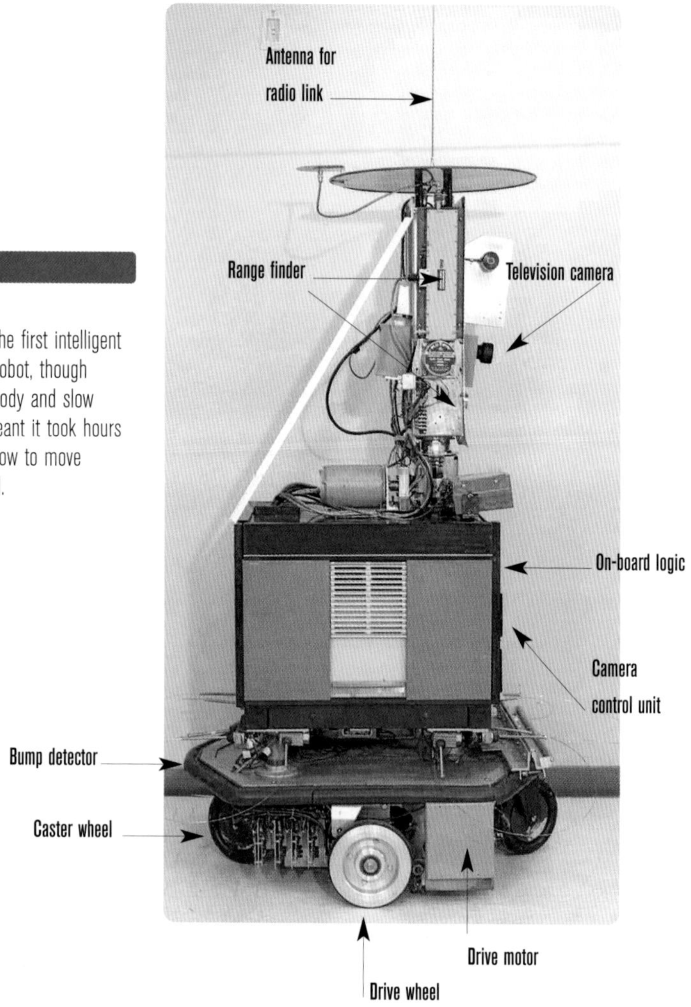

Antenna for
radio link

Range finder

Television camera

On-board logic

Camera
control unit

Bump detector

Caster wheel

Drive motor

Drive wheel

RIGHT

SHAKEY

Shakey was the first intelligent autonomous robot, though its primitive body and slow computers meant it took hours to work out how to move blocks around.

WE HAVE SEEN THAT ROBOTS CAN decide what to do without planning in the human sense. In fact, when it comes to moving and sensing, planning is probably the wrong thing to do—do you plan where to put your legs when you walk? Still, the essence of planning is deciding what to do ahead of time, and sometimes that is the only way to solve problems and avoid difficulties.

Shakey the robot was a research project carried out between 1968 and 1974 by the Stanford Research Institute in the United States (now called SRI Technology) that tried to incorporate planning. It used a version of the three-layer architecture we looked at earlier. Low-level action routines dealt with moving, turning, and planning routes. Intermediate-level actions consisted of sequences of low-level ones and accomplished more complex tasks. At the highest level, a user set Shakey goals, for which it would plan a set of actions and then execute them. These plans were generalized and saved for possible future use.

Shakey wasn't a huge functional success—it spent most of its time sitting and thinking, connected to two large computers by radio and video links, and took many hours to complete simple block-moving tasks. But it did have a huge influence on robot research, and the approach taken by its planning system—the Stanford Research Institute Problem Solver, or STRIPS—is still used today for some applications.

A CUNNING PLAN

It's hard to think of a robot as "intelligent" if its capabilities extend only to those of a cockroach or a termite, even if those are a good deal more difficult to mimic than we thought. This must be why some of the earliest work in robotics, back in the 1960s and 70s, tried to produce robots that could make plans. After all, what could be more characteristic of human intelligence than the ability to plan ahead?

PLANNING WITH STRIPS

Using STRIPS, any action could be represented in an apparently simple way. First, there would be a set of facts that had to be true for the action to make sense. For example, if a robot was going to pick up a cup from a desk, it had to be positioned by the desk. Then there would be a set of changes expressing the results of the action: some new facts would become true (the ADD list), and some facts would stop being true (the DELETE list). So when the robot had picked up the cup, a new fact—that the cup was held by the robot—became true, while the fact that the cup was on the table became false.

The important thing about this way of representing an action is that it is generic—the action has parameters that can be filled in when it is selected during planning. This means that the same pickup action could be used for a plan to pick up any cup in the building, or indeed for other objects that can be fitted into its thing-to-be-picked-up slot. The preconditions and the ADD and DELETE lists can be represented in the same generic way so that the pickup action can be used for picking up a variety of objects in a variety of locations. It is a way of holding some abstract logical knowledge about what is required to pick

something up, and what happens when you do. To make a plan, of course, a number of actions have to be put together in the correct order. There are two basic ways of doing this. The first is to start from the current situation, look at the available actions and choose one that seems to take you toward where you want to be. So if you are trying to make a cup of coffee, the action that takes the cup out of the cupboard would be a good initial move.

Less intuitive from a human perspective, but used in many planning systems, is to start from where you want to be and work backward. So something like: "let's assume I have my cup of coffee; what action must I have carried out—try *pour in milk*." Then you'd insert the action before that—say, *pour in boiling water*.

Arguably, the trouble with robot planning systems is that they try to use a logical mechanism like this. Humans can make plans from first principles in this way but frequently apply lots of other knowledge about how things work, past experiences, things they've read about or been told, analogies and metaphors. Robot planning works well on simple but not very useful tasks, like constructing towers out of building blocks, but still struggles with the kind of planning that gets us through our working day.

BELOW LEFT

START HERE
Planning starts from a set of facts known to be true.

BELOW RIGHT

PUTTING A BLOCK DOWN
It is possible to calculate the state of the world after an action has taken place: an action is represented by a set of preconditions—what must be true for the action stack to be usable—and a set of ADDS and DELETES saying what new facts will become true and which existing ones will become false.

The state of the world before a block is put down.

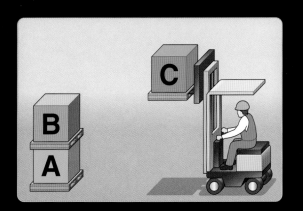

What must be true for a block to be put down.

What changes after a block has been put down.

The state of the world after the block has been put down.

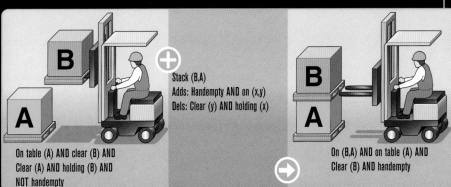

On table (A) AND clear (B) AND Clear (A) AND holding (B) AND NOT handempty

Stack (B,A)
Adds: Handempty AND on (x,y)
Dels: Clear (y) AND holding (x)

On (B,A) AND on table (A) AND Clear (B) AND handempty

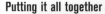

COGITO ERGO

We've seen that recent robotics research has placed an emphasis on the lower-level capabilities humans share with other animals. Some research groups have reacted against this by starting up an area they call cognitive robotics, which the group at the University of Toronto in Canada describe as being about "higher level cognitive functions that involve reasoning, for example, about goals, perception, actions, the mental states of other agents, and collaborative task execution."

ABOVE

ENTIRELY LOGICAL
Aristotle, the Greek thinker (384-322 B.C.), is considered the father of logic.

THE IDEA IS THAT AN INTELLIGENT ROBOT should be driven by logic, and researchers in cognitive robotics are putting a lot of effort into constructing logic-programing languages that can be used to program such robots. This raises an interesting question—should we be trying to construct a robot using what we know about the way humans work, or should we try and construct one that is in some way better than humans?

Psychologists have discovered experimentally what many of us must have long suspected—logic is not wired into the human brain, but is something we have created. An experiment carried out in the early 1970s by a couple of British psychologists showed that the human ability to reason with logic depends heavily on the specific subject matter they are asked to reason with. In this experiment, four cards were laid out on a table bearing the following symbols: **E K 4 7**

The subjects knew that all the cards had a letter on one side and a number on the other, and they were given the following rule:

"If a card has a vowel on one side then it has an even number on the other."

They were asked to choose which cards to turn over to prove whether this rule was true or not. Which cards would you turn over? (The answer is on the opposite page.) The experiment showed that few people were able to reason correctly about which cards to turn over.

However, when the experiment was repeated with cards representing journeys, with a city on one side and a means of transport on the other, the success rate shot up. For example, cards could be: Paris, Madrid, plane, train and the rule to be tested would be: "Every time I go to Paris I travel by plane."

Which cards would you turn over to prove or disprove this rule? (The answer is on the right).

If logic were wired into the brain, it should make no

RIGHT

AIBO ROBOCUP

It takes a robot a long time to process logic—a bit tricky in a spontaneous game!

BACKGROUND

BRAINPOWER

Logic does not appear to be wired into the brain.

ANSWERS

Of course you have to turn over the "E" as it is a vowel and, if the rule is true, it should have an even number on the reverse. This is clear to nearly everyone. You also have to turn over the "7" as it is an odd number and if it has a vowel on the back this disproves the rule. Only twelve percent of people in the original experiment realized this. In the same way you have to turn "Paris" over to see if it has "plane" on the back, and "train" over to see if it has "Paris" on the back. Sixty percent of participants in the experiment realized the second card had to be turned over as well.

difference whether the rule is presented abstractly with numbers and letters, or concretely as towns and modes of transportation. The fact that it does suggests that humans are actually using knowledge about how the real world works—that is, experience—and not logic.

MORE THAN HUMAN?

Just because we don't have logic wired into our brains, does it mean that robots shouldn't have either? Maybe the world would be a better place if humans were more logical? Perhaps we should make robots that are?

Logic reasons with symbols, like the letters and numbers above. One problem for robots would be that they don't get symbols from their sensors. Turning what they do get into symbols would take a lot of time and computation and is prone to error. Another problem is that the propositional logic we use to solve this problem, one that goes all the way back to Aristotle and the ancient

Greeks, assumes a fixed set of symbols. But in the real world this isn't true: drop a vase on a stone floor and you no longer have a vase, but a pile of broken china that didn't exist before you dropped the vase. Researchers have made new logics that can cope with time and change, but these turn out to be very hard to reason with.

Yet another problem we'd face is the principle of logical omniscience. This says that whenever you learn a new fact, you should make every possible inference from it. But in the real world where we learn new things all the time, this would mean comparing every new fact with every existing one, and this would very quickly get out of hand. Maybe researchers will find ways round these problems eventually, but just now a logical robot that can cope with the real world looks a long way off.

Dancing Queen

THE ROBOCUP INITIATIVE HAS AS ITS OVERRIDING aim the fostering of robotics in general. So involving the engineers and scientists of tomorrow has been important right from the start. 1998 saw the inauguration of RoboCupJunior, with a soccer demonstration at the RoboCup 98 event in Paris. The following year, the first interactive workshops were held at the RoboCup 99 event in Stockholm, and the first ever international Junior tournament took place in Melbourne, Australia at RoboCup 2000. Over 100 children from 25 schools in Australia, Germany, and the United States participated in this event.

As we have seen, getting robots to play soccer, even with simplified rules, is not at all easy, and the skills required mean that RoboCupJunior teams are normally run by teenage students. To encourage the involvement of younger children, RoboCupJunior created a Dance Challenge, open to students under the age of 12, where controlling the motor action of a single robot is enough and sensor reading and response is not necessary.

In the Dance Challenge, students build robots that perform to music for up to two minutes. There are no restrictions on the size or number of robots allowed. Emphasis is placed on creativity: some students dance alongside their robots; some have told stories while their robots move; others stand on the sidelines and let their robots perform alone.

Grace and Max Petre, two young students from the United Kingdom, entered the 2000 RoboCopJunior dance challenge, with a robot they had designed and built themselves using the popular LEGO Mindstorms robot kit. Their mother, Marion Petre, says:

"It uses two Mindstorms bricks and six motors—a 'shoulder' section moves independently of the 'pelvis,' wings flap independently, and the head rotates. The two bricks are coordinated with a start switch (a touch sensor). The way the children describe their roles is that Grace was the designer (she decided on things like the wings and wiggles and choreography, and she fashioned the head), while Max was the engineer (he sorted out mechanics and programming).

"The robot is called 'King of Turtles.' It danced to Shania Twain's *That Don't Impress Me Much* (Grace's choice). They did a few weeks of experimenting, and then a week or two of serious designing and debugging. Max in particular made all sorts of engineering and programming insights, like how gearing works, the relationship between power and control, and why programming constructs such as variables exist. We came up with a saying—'It's not the false step that matters, but the next step'—which is fundamental to good engineering practice."

ROBOTS AND EDUCATION

As part of the RoboCupJunior 2000 event, a small team of academics from the United States, the United Kingdom, and Denmark, interviewed some of the teachers who had traveled there with their teams. They wanted to assess whether entering a team was just good fun, or whether there was evidence of real educational gains. Their report was cautious—only 12 teachers had been interviewed and, as they pointed out, this was a self-selecting sample of people who had made the effort to bring a team, and so were probably positive to start with.

These reservations aside, the teachers saw the experience as a very positive one for their students. Having a deadline was very motivating and mixing with adult researchers who were competing in the RoboCup at the same time provided good role models. In addition, having to work together in teams taught students about cooperation, sharing out work, and responsibility, as well as problem-solving, mechanics, and electronics.

One downside was that less than 10 percent of the participating students were girls, and these were all younger students involved in the Dance Challenge. Teachers weren't sure whether this was because dance was more attractive to female students than soccer, or because younger students didn't have such stereotyped views about robotics. One teacher of the older age groups—who taught in a school where robotics was part of the curriculum—commented that female students there saw it very much as a boys' activity.

RoboCupJunior is by no means the only robotics event to involve school students in robotics. The international RoboFesta movement (www.robofesta.net; www.robofesta-europe.org) organizes events with a variety of robot games, while the First LEGO League, for schools using the LEGO Mindstorms kit, has involved some tens of thousands of children in the United States, Singapore, China, Norway, and Denmark. Robots are not just for adult researchers.

HOBBYBOTS

Immense efforts have been expended on creating better methods of controlling robots—on what, if they were living things, we would call their brain. Quite a bit of time has gone into developing sensors, and wheels, legs, wings, and other actuators. The robot body, on the other hand, is rarely more than a lump of metal connecting together the interesting bits. And yet in the natural world, bodies are not at all like this. They are not made of metal, nor even—in most cases—of hard, rigid substances. Why then are robot bodies the way they are?

One explanation is to make them easier to construct: the simplest possible mobile robot is really little more than a metal box on wheels with a computer inside. Another is because in our imagination we still connect robots with moving statues, mechanical warriors with superhuman strength. We think of them as tough metal machines, not as fragile biological mechanisms. But if metal bodies were useful in our everyday lives we would all walk around wearing armored suits. We don't do this for a number of reasons.

First, armor is expensive to make and requires a whole series of technologies to produce. Second, we have made ourselves safe environments containing few obvious threats to our physical safety. Third, metal is heavy and uncomfortable: it gets in the way of necessary bodily processes like sweating; it dulls our sense of touch; and it requires a lot of extra energy to carry around. Because it is inflexible, even as chain mail, it also impedes movement and gets in the way of everyday activity. It is for these last reasons that few animals have evolved massively armored bodies, and that in particular animals with shells are small and relatively immobile. Metal is much less perishable than flesh, but it has no ability to repair itself or to regenerate. When the bodywork on your car rusts through, it needs welding, not time to heal.

BACKGROUND

MUSCLES IN MOTION
The cheetah's body is a lot more than a convenient way of connecting its legs to its brain. Flexible and self-healing, it maintains the correct internal temperature, circulates nutrients to the muscles and provides the oxygen which powers its phenomenal sprint.

Chapter 5

SHAME ABOUT THE BODY...

DESIGN DECISIONS

We have seen that biomechanics—the study of the mechanics of living things—has contributed many fine ideas to robotics. But some biomechanics is tied very closely to the way animals' bodies are made, of flesh and blood, not of metal. Once you have a heavy metal body this influences a lot of other design decisions: to move large lumps of metal, for example, you need motors, and these function completely differently from the muscles of mammalian bodies. But muscles have a number of advantages over motors, and roboticists are now starting to investigate them more closely. If you rely on motors, you need power to drive them, and that nearly always means electricity. As we will see, the limitations imposed by this requirement present a formidable obstacle to the development of autonomous robots. Animals, on the other hand, although they use some electricity, are basically driven by chemical power.

The rigidity of metal makes it difficult to incorporate any equivalent of touch in a robot. A sense of touch requires compliance—a surface has to give a little when it comes into contact with something in order to transmit some of the force it experiences back into the robot—and metal is not compliant. This is one reason why the study of haptics—force-feedback and touch—is so much less developed than that of the sensor equivalents of vision or even hearing.

As a result, robots are relatively dangerous around humans. Heavy metal bodies with no sense of touch could cause a lot of damage—compare the effect of a fellow human standing on your toe in a crowd and a metal robot of the same dimensions running over your foot. And being jostled by other people is less of a problem than being squashed by a full-size metal body that can't feel you're there.

RIGHT

STIFF
Robots are typically large metal constructions: heavy, inflexible, and lots of work to move.

ABOVE

BENDY
Human bodies are flexible, able to repair themselves, and covered with touch sensors.

US AND THEM

The metal body of a robot doesn't much resemble the biological tissues of an animal, and not simply because it is made of different materials. The animal's body and brain form an organic whole, full of processes that regulate and maintain. The robot's body is just a container for the electronics, and the motors and gears that drive its wheels. Whatever processes take place within its metal body—rusting or corrosion, for example—are quite outside the robot's control, and are not really part of it at all.

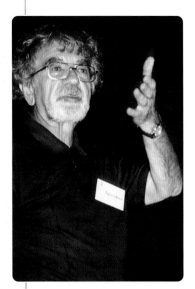

HUMBERTO
MATURANA
Maturana and Varela defined the important idea of autopoiesis in living organisms.

IN THE SECOND HALF OF THE TWENTIETH century, Chilean biologists Humberto Maturana and Francisco Varela tried to take a very general view of the way biological organisms worked. They pointed out that the whole of a living creature, whether a single-celled organism or a large and complicated mammal, forms a homeostatic system. This is one in which critical variables are kept within defined limits, much as the thermostat on an airconditioning unit turns the airconditioning on or off to keep the room temperature within certain limits. Sweating is one of the processes which helps to keep your body temperature within acceptable limits. If your temperature were allowed to go too high you could die of a fever; too low and you could die of hypothermia.

Maturana and Varela argued that living organisms were different from other kinds of homeostats in that, rather than just keeping certain values within limits, they try to maintain their own internal organizations and structures, and all the processes that go along with that. While the airconditioning thermostat just maintains the air temperature in a room, you maintain your body temperature because the processes that make up life in the body are disrupted outside of certain limits. Because organisms can adapt and grow, the physical properties of their systems may change over time—think of the difference between the body of a baby and that of an adult. This doesn't matter as long as the overall organizational structure is maintained. They called this special sort of homeostasis *autopoiesis*.

Maturana and Varela also turned their idea around. Not only, they suggested, were living organisms autopoietic machines, but any physical system that demonstrated autopoiesis must be living. The implications of this for robotics are clear: a living robot is one that can maintain all its internal processes in a changing environment.

PROCESSES—WHAT PROCESSES?

If we look at the human body from a functional point of view, we see that there are a whole set of interlinked processes at work. The central nervous system can be thought of as the brain distributed out around the body—sending control messages from the brain to the organs and limbs, and feeding back data on how the body is functioning. Pain allows the brain to know if damage is being sustained; the proprioperceptive sensing we considered in Chapter 3 is achieved by passing back information on the actions performed by the muscles to that part of the brain that keeps track of where the arms and legs are, and what they are doing.

The circulatory system takes oxygen from the air via the lungs, and also nutrients from the digestive system, into the blood; both are transported to the muscles, where they are chemically combined to produce the energy that the muscles need to function. The blood also removes the waste products of this reaction. The digestive system takes food from the outside environment and processes it into a form that can be carried in the blood, as well as excreting the parts that can't be used.

The endocrine system uses glands and hormones to regulate most of the other systems in the body—the physiological changes of the "fight or flight" reflex, for example, that result from a release of adrenaline, and that affect heart rate, circulation, and sense of time, among other things. Growth and the physical changes experienced in adolescence are also down to the endocrine system. The reproductive system supports mating behavior, and in females the complex process of producing offspring.

These processes are all interlinked, and autopoiesis is the overall process that keeps them all going in the right sort of interlinked way for as long as possible. When this is no longer possible the organism dies, but if it has successfully reproduced, a similar set of processes will have been inherited by its offspring.

So what would it mean in practice for a robot to have these kinds of processes? If it had them, would we agree that it was alive? And would it then be subject to death just like us?

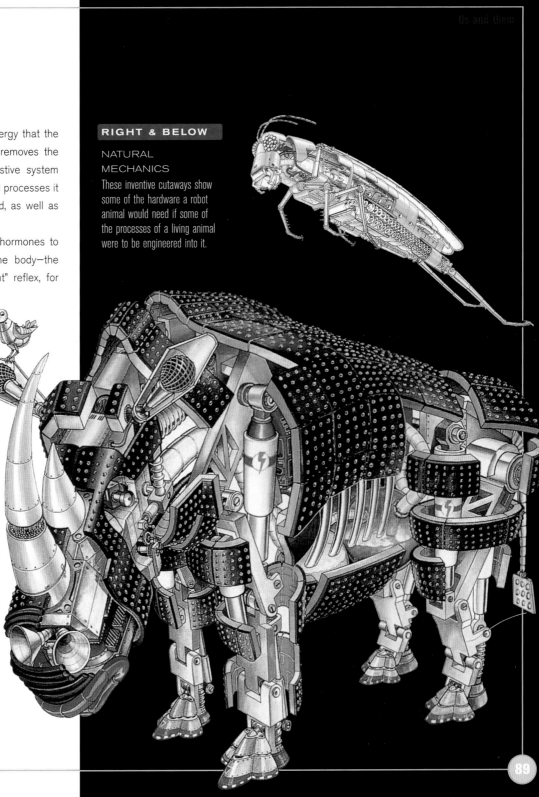

RIGHT & BELOW

NATURAL MECHANICS
These inventive cutaways show some of the hardware a robot animal would need if some of the processes of a living animal were to be engineered into it.

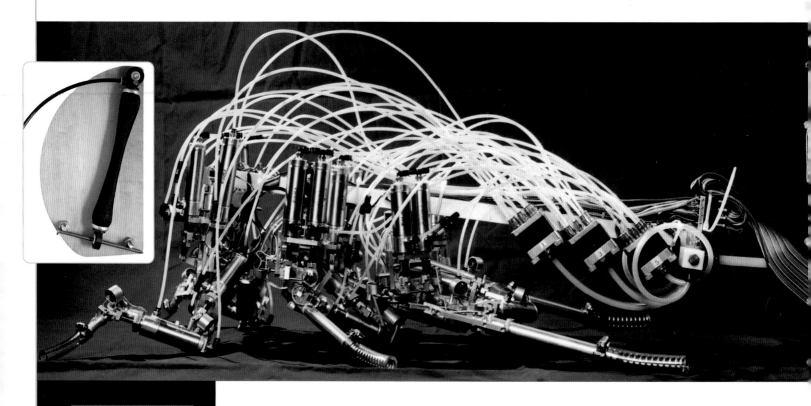

ABOVE

PUMPING AIR
A McKibben artificial muscle is driven pneumatically. As it contracts it produces a lot of force for its size and weight.

ABOVE RIGHT

MUSCLEBOUND
The most recent robot cockroach from Case Western Reserve University uses McKibben artificial muscles.

SHOW US YOUR MUSCLES

In contrast to living creatures, most robots move with the aid of electric motors. These motors are small and relatively inexpensive, and work best when they can drive wheels at fairly constant speeds. Maximum power is achieved when they are turning at one specific (ideal) speed and their top speed—with no load—is only twice this ideal. If a robot needs a lower speed to carry out some task that must be performed more slowly then a gearbox is required to gear the motor down. But without variable ratio gearing the robot is then condemned to do everything at this speed.

THIS BECOMES MORE OF A PROBLEM when robots are not just running about on wheels—when they have legs, for example, or an arm that has to pick up objects of very different weights. Gear reduction makes the motor rather intolerant of shocks, such as might be caused by a leg hitting the ground. Stepping motors can be used, but these are big, heavy, expensive, power-hungry, and low in efficiency. Motors are also inclined to burn out if overloaded, at which point they are no use at all.

Mammals use muscles for movement, fixed to a rigid skeleton so that the bones move like levers. Muscle fibers contract and expand, and this allows a pair of opposed muscles—an antagonistic pair—to control each major limb, with the fibers on one muscle contracting and the fibers of the other expanding at the same time. Research has shown that our muscles are made of two types of fibers: "fast" fibers for quick movements, and "slow" fibers for slow and strong movements. To produce artificial muscles, some way of controlling expansion and contraction is needed, and researchers have been looking at ways of doing this.

Two solutions are available off the shelf. One is the McKibben Artificial Muscle, driven pneumatically, and first developed for use in artificial limbs in the 1950s. This was commercialized in the 1980s by the Bridgestone Rubber Company of Japan and in the 1990s by the Shadow Robot Group in England for robotic applications.

An expandable elastic tube, or bladder, is surrounded by a braided shell. When the internal bladder is pressurized, it expands like a balloon against the braided shell, which constrains the expansion to keep the cylindrical shape. As the pressure increases, the whole thing shortens just as a real muscle would. When the pressure is released, it resumes its original shape and length. McKibben muscles can be constructed or bought in a variety of sizes and have a very high power-to-weight ratio, particularly useful for mobile robots. The only tricky aspect is providing the compressed air for the pneumatics, but a number of robots have already used these devices, from the most recent Case Western Reserve University cockroach-based hexapod to the anthroform arm being built at Carnegie Mellon University, both in the United States.

LIVE WIRE

A second quite different approach uses Shape Memory Alloy (SMA). Some years ago, researchers discovered that certain metallic alloys, when formed into a shape at a particular temperature, would reassume the same shape whenever they were returned to that temperature as if they "remembered" it. One of the most widely used is a nickel-titanium alloy called Nitinol, which when heated by an electric current will compress by up to eight percent.

The Northwestern University Robot Lobster (see pages 50-51) uses this material: a single strand or loop of Nitinol insulated from seawater by a Teflon sheath. When a series of current pulses heat the Nitinol, it contracts. When the pulses cease, the surrounding water cools the material and it expands again. These muscle wires are very strong indeed and can lift thousands of times their own weight. The downside is that the contraction isn't really substantial enough for anything other than fairly small movements, and the wire is very expensive.

Researchers at NASA's Jet Propulsion Laboratory in California have been trying something similar, but this time with a group of plastics called electroactive polymers (EAPs). They created a ribbon made of chains of carbon, fluorine, and oxygen molecules. When an electric charge flows through the ribbon, charged particles in the polymer get pushed or pulled on the ribbon's two sides, depending on their polarity and on the ribbon bends. Four ribbons were put together to make a gripper strong enough to pick up a rock. Other novel materials are also on the way—muscular robots are no longer a dream.

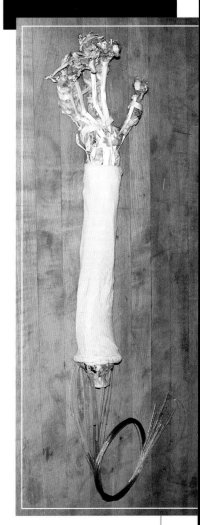

BELOW

ANTHROFORM ARM
This arm and hand may not look very human, but they form an accurate biological model, and the McKibben muscles match the human ones.

A SOFT TOUCH

There is a lot more to human skin than meets the eye. Of course, it protects our insides from what's on the outside—including pressure, heat, cold, ultraviolet rays from the sun, chemical agents, and pathogens. It is also a vital part of the human thermoregulation system. The subcutaneous fatty tissue acts as insulation against the cold, while the skin cools down the body by releasing water during the sweating process.

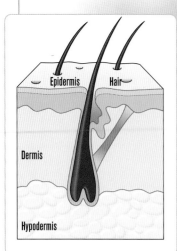

Epidermis Hair

Dermis

Hypodermis

SKIN IS ALSO FULL OF SENSORY CELLS, AND SUPPLIES US WITH a great deal of information about temperature, local damage, and pressure. We touch and sense our environment through the skin. It may be much less hardwearing than metal, but when functioning normally, skin renews itself every 28 days: dead cells fall off and new ones grow from below. All this makes it a complex structure, with blood vessels, nerves, pores, sweat glands, and hair follicles, as well as a set of layers.

Skin also reflects the internal state of the organism to some extent: think of flushing, which reflects hormonal changes below the skin. There's the funny white color people go around the mouth when they feel nauseous, the redness of fever, or the gray color of heavy smokers whose circulation has been impaired by the resulting narrowing of blood vessels.

Much work has been done on artificial skin, most of it for humans who have suffered major burns. The idea is to grow new skin, often using a polymer base to help the structure and a layer of silicon to protect it until the individual's own skin grows through. This is of limited use to a robot, and of course many of the functions of skin would only be useful for a robot that had similar internal processes to a human.

The basic protective function, though, is something that is needed for underwater robots— we have already seen (on pages 32–33) how robot pike and tuna are equipped with a latex skin to keep the electronics dry. Robot snakes, whose bodies are in more-or-less permanent contact with the ground, but also have to flex to move, also need a plastic or rubber skin. Up to now, the idea of a skin for ordinary landbased robots hasn't seemed necessary except for the sake of appearances—that is, to make a humanoid robot look more human. As we'll see in the next chapter, though, making a robot look human is not automatically a good idea, because of the reactions it provokes in real people.

ADDING TACTILE RESPONSE

The really useful function of skin for most robots would be to give them an equivalent to our sense of touch. To add touch, a robot body would have to be surrounded by a web of tactile sensors. Embedding them in a skin would be a good way of keeping them all together and in the correct position. The most popular approach to tactile sensing uses the piezoelectric effect, in which mechanical energy—pressure—is converted to electrical energy. It was discovered in the 1880s by the Curie brothers, who found that when pressure (piezo means pressure in Greek) was applied to a polarized crystal, the consequent bending resulted in an electrical charge. A sensor can be made by sandwiching the crystal between two metal plates, making a capacitor. The more force is applied to it, the greater the charge, and this can be measured and used to control the robot's behavior like any other sensor.

Touch sensors are usually attached as small pads, on robot grippers, for example, where the ability to feel the object being picked up is crucial if it is not to be squashed by the robot grasp. Making arrays of sensors and embedding them in an elastic material gives something a little bit like skin, but there are still many problems to be solved. For example, though it's easy to calculate what electrical charge a known pressure on such a "skin" will produce, because of the elasticity of the material used, a given electrical charge may be produced by a number of different patterns of pressure, so that it's hard to use this "skin" as a reliable pressure sensor. Some of the learning technologies we looked at in the last chapter might help to deal with this. But human touch detects much more than just pressure—think of stickiness, texture, friction, hardness, and elasticity—and all these have yet to be tackled.

Work by a team at the University of Illinois in the United States has produced a novel material that can actually repair itself, though to date this is intended for aircraft wings rather than robots. Many artefacts—including some wings—are made of a composite polymer, in which small fibers of materials, such as glass or carbon, are embedded in a polymer matrix. These composites can be cracked and permanently damaged by vibration and bending under loads. The Illinois self-healing composite contains tiny capsules of polymer monomers (the building blocks from which polymers are made) with catalysts that cause polymerization. Cracks in this material rupture some of the embedded capsules, releasing the monomers, which link up with each other as they come in contact with the catalyst and bond the fracture faces together.

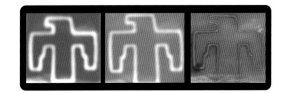

SKIN COLOR
NASA scientists have produced a plastic skin which changes color in different conditions, allowing whatever is covered in it to reflect what is happening in its environment.

BELOW

LAYERS OF SKIN
The multiple functions of skin are reflected in its complex structure. This complexity and functionality would be extremely difficult to build into any robot equivalent of skin.

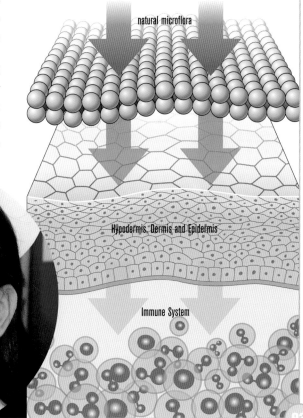

natural microflora

Hypodermis, Dermis and Epidermis

Immune System

RIGHT

FACING FACTS
Skin can be much more than mere protection for a robot. Hiroshi Kobayashi in Japan has used a plastic skin on his Face Robots to make them look more human, and so they can produce facial expressions in the same sort of way as a human would.

WHEN THE BATTERY IS DEAD, THEN THE robot to which it belongs is dead too. But even before this happens, the drop in voltage as a battery runs down can result in erratic behavior as the robot's electronics try to cope with the declining power supply. The obvious answer is to enable a robot to sense when its battery is running low—a sort of "hunger" sensor—and give it the ability to find somewhere to recharge itself. First, then, the robot has to use rechargeable batteries.

Batteries work via electrochemical reactions between a positive and a negative electrode. When supplying power, an oxidizing reaction takes place at the negative electrode, and an opposite reducing reaction at the positive electrode. To recharge a battery, an electric current is applied in the opposite direction, reversing the power-generating reactions and returning the electrode materials to their original physical state. Unfortunately, some electrochemical reactions don't reverse as neatly as this.

For example, if you try to recharge the carbon-fluoride-lithium batteries often used in cameras, the cell electrolyte decomposes, and eventually the fluoride is oxidized to form fluorine gas. In other cases, the reaction reverses, but the original metal reappears in lumps after repeated charge-discharge cycles. This eventually forms an electrode so big it meets the other one and a short circuit occurs. Until recently it was not possible to recharge an ordinary alkaline battery (which uses manganese and zinc): not only did the battery perform less efficiently with each recharge, but during recharging it might generate enough hydrogen gas to cause a small explosion.

Alkaline batteries have been redesigned to get around these problems, by including, among other things, safety vents to prevent the buildup of excess pressure during

BATTERIES NOT INCLUDED

The biggest single obstacle to the development of autonomous robots is not mobility, intelligence, or sensing, however hard those might be to perfect: it is the limitations of the power source. The wonderfully engineered Honda humanoid robots can keep going for about 30 minutes on the battery pack they carry on their back. The much smaller Sony Aibo robot dogs have a battery that lasts for just two hours. If robots are going to take over the world, they'll have to do it pretty fast before their batteries run out.

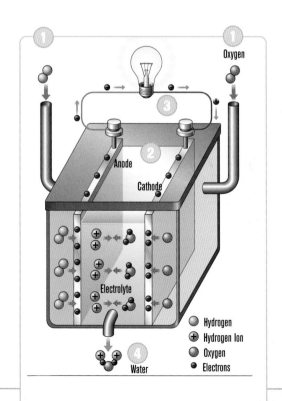

recharging. Nickel-cadmium (NiCad) batteries convert back to their original electrode materials in a smooth, compact form when recharged.

GETTING CHARGED UP

The robot with a rechargeable battery still has to find somewhere to recharge. Ordinary wall sockets were not designed for robot access, so researchers often create a sort of recharging hutch, into which a wheeled robot can roll and make easy contact with a mains supply. But of course the robot has to be able to find the recharging station when its charge is low, and this involves both sensor activity and some sensible choice about when to abandon whatever else it is doing given the likely effort involved in getting to a station. Then if a set of robots is sharing such a station, there's the problem of forming an orderly line and not going dead while waiting. The limited capacity of even the best batteries currently available makes all this quite a challenge.

Since batteries are also typically large and heavy, other portable sources of electricity are being researched. Fuel cells are twice as efficient as batteries and do not need recharging—they use a reaction between hydrogen and oxygen from the air to generate an electrical current which continues for as long as hydrogen fuel is available. The hydrogen can come from the hydrocarbon natural gas, but wastewater digesters or land-fills are also possible sources. One leading design is the PEM, or Proton Exchange Membrane. This uses a platinum catalyst on the surface of the mem-brane where hydrogen and oxygen combine to form water and electrical energy.

This technology is also being actively researched as an alternative to the internal combustion engine for cars and buses, and the problem of refueling for robots may be solved if it becomes widely used. As we'll see, though, living things manage energy differently, and this may provide a quite different way of energizing a robot.

ENERGIZING

Animals and robots manage their energy requirements in vastly contrasting ways. Robots drive motors using electricity. Animals operate muscles using chemical power. When a battery dies, the robot is also dead. If an animal misses a meal, fat deposits allow it to continue for days or even weeks, and fat is continually laid down whenever an animal consumes more food than its immediate needs demand. In dire emergencies, even the animal's muscles can be consumed as fuel—whereas the robot's metal body is a dead weight that it has to move whatever its energy resources.

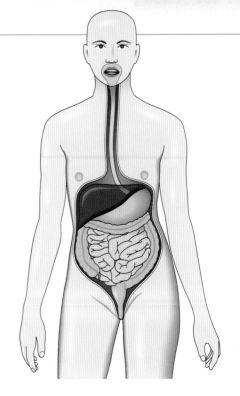

THE ANIMAL METABOLISM IS ALSO MUCH MORE FLEXIBLE—YOUR THYROID GLAND CAN turn your metabolism up or down depending on circumstances, while your muscles can consume fuel both aerobically, using oxygen, or if not enough oxygen is available, anaerobically for a limited period, making up the "oxygen debt" later. Finally, an animal's waste products are relatively easy to dispose of—carbon dioxide and water are the direct result of supplying energy to the muscles, while solid waste and uric acid are the byproducts disposed of by the digestive process that produces the sugar for the muscles. Dead batteries present much more of a disposal problem.

BELOW

ENTOMOPTER
The Georgia Tech entomopter uses chemical energy to flap its small wings fast enough to keep it in the air.

THE MICROBIAL FUEL CELL

A really autonomous robot ought to be able to "live off the country" in the way an animal can. Assuming that the robot is to be powered by electricity, what is needed is a method of turning the carbohydrates found in vegetation into electrical power. The microbial fuel cell (MFC) arising from the work of the bioelectro-chemistry group at King's College, London, seems like a promising way of doing this. The basic idea involved is to use the action of micro-organisms—bacteria and yeast have both been tried—on carbohydrate or sugar to release energy. Under ideal conditions an MFC can run on only food, water, and air, and in some experiments they have succeeded in operating continuously for at least three months.

Making an MFC that can run a robot is a much more challenging problem. A team at the University of South Florida is building what it calls a "gastrobot," a robot driven by an MFC in such a way that it can run indefinitely on just food, water, and air. In this sort of system, the MFC is just one of many components: the whole cycle of foraging, harvesting, mastication, digestion, and defecation has to be tackled.

ABOVE RIGHT

GASTROBOT

Stuart Wilkinson from the University of South Florida feeds a sugar lump to the prototype gastrobot. A stack of six individual MFCs is located in the center wagon. The other wagons contain anolyte (anode and electrolyte) and catholyte (cathode and electrolyte) tanks along with gastric and heart pumps. A DC motor drives the robot's wheels.

ABOVE LEFT

DIGESTION

The human digestive system breaks down food into usable proteins, minerals, carbohydrates, fats, and other substances. Three chemical reactions, involving specific enzymes, take place: the conversion of carbohydrates into simple sugars such as glucose; the breaking down of proteins into amino acids such as alanine; and the conversion of fats into fatty acids and glycerol.

The list of research challenges here is long. For example, how does the gastrobot know when its MFC needs more fuel? Eating too much actually makes the MFC less efficient, and at worst might clog it up—the gastrobot needs an appetite equivalent, to know when feeding is required. Monitoring the amount of carbon dioxide produced by the fermentation process, a sort of gastrobot flatulence, is one possibility. A further problem is that many of the micro-organisms that might work in MFCs are not good for humans at all, raising safety issues. Some of the most effective turn out—unsurprisingly—to be organisms like *E. coli* that live in the human gut and work very well there. Ingesting *E. coli* can give humans nasty gastrointestinal or blood infections.

South Florida has developed an experimental gastrobot to research these problems. Its power is generated by a stack of MFCs, whose "food" currently consists of sugar lumps fed manually into its "stomach" tank. Constructed like a small train to carry the various bits of digestive apparatus, it looks unlike most other robots, but the ideas it embodies may mark an important step forward.

RECIPROCAL CHEMICAL MUSCLE

Another group using chemical power—in this case directly, without making electricity—is working on a micro-flyer at Georgia Tech in the United States. Robert Michelson has designed what he calls a Reciprocating Chemical Muscle as a way of driving the entomopter—a small insectlike flying robot. A chemical propellant is injected into the body, causing a reaction that releases a gas. The gas pressure operates a piston in the fuselage that is linked to the wings, causing a rapid flapping motion. Vents in the wings allow the gas to be discharged, changing the lift on either wing to produce a turn. Michelson believes the technology will work on a much smaller scale, and hopes eventually to shrink the entomopter down to the scale of living insects.

SMALL IS BEAUTIFUL

We have seen that the most successful complex living creatures around are the insects, to judge by the number of species. One characteristic of insects we haven't yet considered is their size: compared to humans they are extremely small. So what are the prospects of us being able to design and manufacture tiny, inexpensive autonomous robots? The idea has enormous attractions and a host of potential applications, from the medical to the military. Is it realistic, or just a daydream?

ABOVE

MICRO-ROBOTS
The United States Department of Energy's Sandia National Laboratories has created possibly the smallest autonomous untethered robot so far. It is ¼ cubic inch (4.1cm³) in size, weighs less than a large walnut, and contains an 8K ROM processor and temperature sensor. Three watch batteries power two motors that drive the tracked wheels, covering about 20 inches (50cm) a minute.

IF OUR STUDY OF BIOMECHANICS MAKES one thing clear, it's that size matters. When a robot is scaled down, say by a factor of ten, its surface area reduces by a factor of 100, and its volume and mass by a factor of 1,000. If the motors that drive it operate at the same speed, it will move ten times more slowly too—its legs would only be one-tenth the length. Its momentum—its mass multiplied by its speed—will be 10,000 times less. And its kinetic energy will be 100,000 times less.

The upshot of all this is that small robots face wildly different problems from those presented to big robots. Where a big robot's momentum might do serious damage to its surroundings were it to hit something with its heavy body, a small robot in a similar collision is more likely to damage itself than anything else. Acceleration forces are a major problem with big robots—with small ones, the mass to be moved is negligible and acceleration is no problem at all. The low momentum also makes small robots behave quite differently—they are able to get over anything they can get a grip on, so obstacle avoidance is no longer such an issue.

The way size makes a difference is very clear in animals. For all the centuries that humans tried to fly like birds, it was never going to be possible: the scaling up of forces from bird size to human size would have required a shoulder joint too massive to operate, just to flap wings of the size needed to lift off. Going in the opposite direction, researchers wondered for a long time how spiders were able to move their legs, given that they had neither muscles nor sinews like larger animals. It turned out that spiders use their own blood as a hydraulic fluid, probably controlling the pressure by using muscles to compress the forepart of their bodies. This isn't as slow and laborious as it sounds, either—some jumping spiders catch their prey so quickly, with leaps of about 20 times their body length, that only video recording can capture exactly how they do it. Many small invertebrates rely on the pressure of internal fluids to keep their bodies firm in the same way that air stabilizes a tire—they have, in effect, a hydrostatic skeleton.

THINKING OUTSIDE THE BOX

This suggests that nanorobots need not necessarily use the same mechanisms of motors and gears as larger robots, which is convenient as making motors and gears on the nanoscale is really very difficult. Novel ways of doing things in small robots is an active area of research—piezoelectric muscles, for example, become much more attractive on these scales. One new technology being investigated is that of nanotubes: cylindrical molecules made of carbon atoms only a millionth of a millimeter in

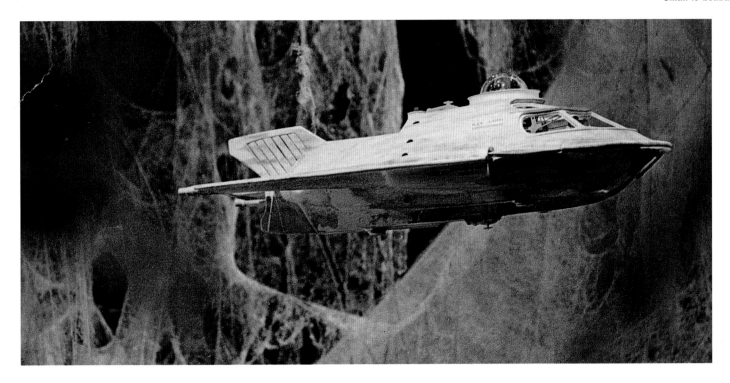

FANTASTIC VOYAGE
The imaginary journey through
the human body portrayed in
Fantastic Voyage, a science
fiction epic of 1966, might
eventually become reality if
nanorobots can be perfected.

diameter. These tiny tubes expand and contract by about one percent along their length when an electrical charge is applied and then removed, making them possible raw material for a nanomuscle.

Ron Fearing and a team at the University of California at Berkeley have been using piezoelectric effects as a way of getting a projected robot fly off the ground. The wing has to both flap and rotate if it is to work like a real fly wing, and do it 150 times a second for a million or so cycles without breaking. Little crystals of piezoelectric ceramic are being used with a complicated kinematic link-age to drive the wing. When a high voltage is applied, the ceramic bends, so that in effect the team have a solid-state motor with no moving parts.

Of course creating a wing structure that weighs only 20 milligrams is not easy—the usual welds and joins are hard to construct and friction is a problem with the small forces involved. Fearing's team has tried folding stainless steel,

starting with a flat sheet of metal and cutting it out with a laser, to the particular pattern they need. Then they fold it up into a hollow stainless steel structure made of very thin sheets, almost like origami.

An artificial fly is still a large robot compared to the vision of a real nanoscale mechanism which could operate on a single molecule. The University of Southern California has set up an interdisciplinary Laboratory for Molecular Robotics that is investigating a possible approach to nanorobot component fabrication, itself using robotic technology to assemble molecular-sized building blocks with scanning probe microscopes. The idea is to position the nanocomponents and then form an assembly using a chemical glue, such as DNA. This work is all at a very preliminary stage so far, but maybe nanorobotics will one day meet genetic engineering.

The Robot Zoo

THERE ARE TWO ALMOST OPPOSITE MOTIVATIONS for biomimetic robotics (robots made using biological principles). We have seen one of them already: the mechanics, physics, and chemistry used by biological organisms are often superior to the designs of engineers and scientists. This should come as no surprise—evolution, by definition, produces very effective adaptations to the environment.

Evolution is not design, though, and it can be difficult to decide which bits of an organism are there because they are effective and which are only there because of the evolutionary path the species has followed. An example in human terms is the appendix, a part of the digestive system with no current useful function, and probably a vestige of the extra digestive apparatus needed to process grass and other vegetation. A robot digestive system based on a human one would not need to include a robot appendix.

The other motivation for pursuing a biomimetic approach is to gain a better understanding of how living things work. Biologists' theories about living creatures can be difficult to test and prove. Engineering the theory into a robot concentrates the mind and forces the biologist to think out all the details. The robot's behavior can then be used to test the theory.

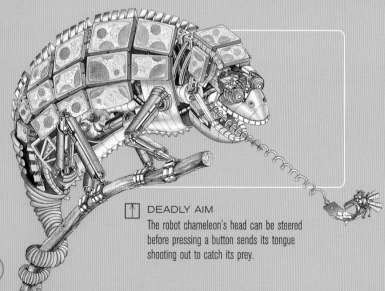

↑ DEADLY AIM
The robot chameleon's head can be steered before pressing a button sends its tongue shooting out to catch its prey.

MECHANICS IN ACTION

But here's a third motivation. Animals make a gripping example of mechanics in action—and an engaging way of teaching the basic principles of biology and mechanics. The Robot Zoo is a result of this third approach. It originally appeared as a wonderful children's book in which a biomechanics expert, Professor Philip Whitfield of King's College, London, teamed up with a talented graphic artist, John Kelly, to portray 16 animals—from the blue whale and the giraffe to the spider and the T4 virus—as if they were robots. In Kelly's extraordinary art, muscles become pistons, intestines become filter tubes, and brains become computers. The idea was then taken up in the United States under the sponsorship of TIME magazine and the computer company Silicon Graphics, and the expert builders of interactive museum exhibits, Clear Channel Entertainment-Exhibitions, Inc., were enrolled to turn the designs into a real robot zoo.

FROM DESIGN TO ROBOT ANIMAL

Eight animals were selected to be turned into real robots: a chameleon, a rhinoceros, a giant squid, a platypus, a housefly with a 10-foot (3-m) wingspread, a grasshopper, a bat, and a giraffe whose head and neck alone measure 9 feet (2.7 m). Some of the animals were to be built at approximately lifesize, while others would be very much scaled-up. We have seen that this can make a difference to the biomechanics, but constructing a fly or a grasshopper to scale is both extremely difficult and would make a poor exhibit in an interactive museum.

Calling the exhibits "robots" is stretching the term a little, as they do not move around autonomously—they are more accurately described as animatronic designs, like those that have been used to produce model dinosaurs in the past. The Robot Zoo exhibits are based on two technologies: compressed air mechanics controlled by interactive 3-dimensional computing systems. Hoses feed compressed air into a bank of computer-controlled valves to operate the movements of some robots. Others have small air compressors within their bodies.

Incorporating computer technology adds a number of interactive activities to the exhibits. Using real-time color image processing on

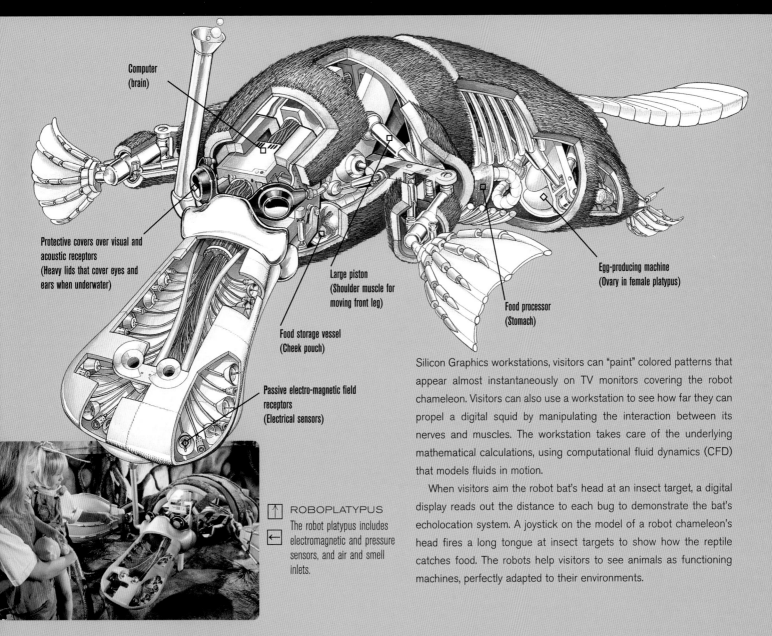

Computer
(brain)

Protective covers over visual and
acoustic receptors
(Heavy lids that cover eyes and
ears when underwater)

Large piston
(Shoulder muscle for
moving front leg)

Egg-producing machine
(Ovary in female platypus)

Food processor
(Stomach)

Food storage vessel
(Cheek pouch)

Passive electro-magnetic field
receptors
(Electrical sensors)

↑
← ROBOPLATYPUS
The robot platypus includes
electromagnetic and pressure
sensors, and air and smell
inlets.

Silicon Graphics workstations, visitors can "paint" colored patterns that
appear almost instantaneously on TV monitors covering the robot
chameleon. Visitors can also use a workstation to see how far they can
propel a digital squid by manipulating the interaction between its
nerves and muscles. The workstation takes care of the underlying
mathematical calculations, using computational fluid dynamics (CFD)
that models fluids in motion.

When visitors aim the robot bat's head at an insect target, a digital
display reads out the distance to each bug to demonstrate the bat's
echolocation system. A joystick on the model of a robot chameleon's
head fires a long tongue at insect targets to show how the reptile
catches food. The robots help visitors to see animals as functioning
machines, perfectly adapted to their environments.

HOBBYBOTS

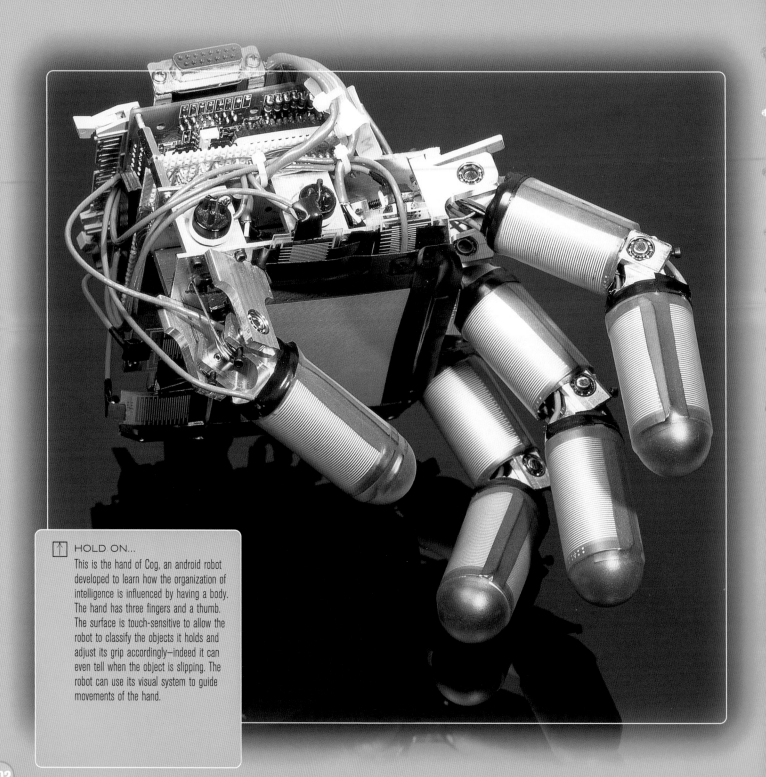

⬆ HOLD ON...

This is the hand of Cog, an android robot developed to learn how the organization of intelligence is influenced by having a body. The hand has three fingers and a thumb. The surface is touch-sensitive to allow the robot to classify the objects it holds and adjust its grip accordingly—indeed it can even tell when the object is slipping. The robot can use its visual system to guide movements of the hand.

POKER FACE

Sarcos, an android entertainment robot, apparently engaged in a game of seven-card stud with a group of robot engineers. Sarcos was built at SARCOS Research Corporation in Salt Lake City to serve as master of ceremonies. Its actions are either pre-programed or run in real time. Mouth movements link with an audio system, while sensors record degrees of body, limb, and finger movements. Paradoxically, it cannot yet play cards!

In the past, the typical robot worked in a car factory—a large manipulator arm with few or no sensors, fixed in one position inside a sturdy fence to stop humans getting in its way. But the big change is just beginning. Robots are moving out of the car factories—and the research labs—into ordinary human environments.

The Sony Aibo robot dog was a trailblazer, turning robots overnight into fashionable adult toys, and it was followed by a large number of entertainment robots. But it isn't just in the sphere of entertainment that robots have made their mark: robot cleaning machines and museum guides have also begun to appear, and in the future we could see robots involved in therapy, in education, and in the home.

"So what?" you say. Robots have been a part of the human world in science fiction since the genre began. Isn't this just a case of the real world catching up with its fictional portrayal? But involving robots in everyday human interaction raises problems that science fiction can conveniently avoid.

A vital feature of human intelligence is the ability to interact with other humans—understanding and following social conventions, imagining the consequences of actions before carrying them out, predicting the behavior and reactions of other humans, and conducting conversations in which each speaker takes his turn.

THEORY OF MIND

Humans develop what is sometimes called a "Theory of Mind," which means we come to understand that other people are autonomous thinking beings with their own points of view. This allows us to experience empathy with other people's problems, and at the same time enables us to deceive them by imagining how our own behavior will appear to them. This ability is normally developed while we are still children, but, curiously, is absent in children suffering from autism. This is probably one of the reasons why people who suffer from autism find social interaction so confusing and stressful—they are not able to put themselves in someone else's shoes

Chapter 6

MAKING FRIENDS

and as a consequence find other people's behavior unpredictable. If robots are to interact successfully with humans, they need to be equipped with some basic guidelines for social intelligence. How to achieve this has become a major focus of research for roboticists.

INTERACTING WITH ROBOTS

Human-computer interaction—HCI—has been studied intensively over the last 20 years as computers have moved out of the machine rooms and into the home—not just into PCs, but into video recorders, washing machines, TV boxes, and, of course, computer games consoles. Understanding the human side of the interaction is vital. We now know, for instance, that the average human short-term memory can usually retain a maximum of seven items before having to memorize them. As a result, computer menus with more than seven items need to be divided up into submenus, otherwise average users will have forgotten the first item in the list by the time they get to the bottom of it.

Are there equivalent human limitations that could help us design robots that interact with humans? We don't yet know, and the field of human-robot interaction—HRI—is still in its beginnings. Should we design our robots to look human? Should they sound human or move like us, or would it be better to make them obviously machinelike? How much social intelligence does a robot need to interact successfully? What feedback does a human have to provide to interact with a robot?

One thing is becoming clear: humans tend to respond to robots as if they were "alive," even if the robots' appearance and capabilities fall well short of those of living things. Their ability to move and sense makes us unconsciously identify robots as animals rather than objects. Robot designers can't wish this away, so they have to learn to take it into account as they tackle the next step.

BACKGROUND

COMPANIONSHIP
Dogs feel like friends, not possessions—could we ever feel like this about a robot?

RIGHT

PETTING THE DOG
A research team from the United States comprising Peter Kahn and Batya Friedman (University of Washington), and Alan Beck and Nancy Edwards (Purdue University) took a Sony Aibo robot to a home for the elderly to see how the residents would react. As these pictures show, even though the Aibo does not look like a real dog, the residents treated it more like an animal than a toy. One pets it as if it were a real dog, while another imitates the way the Aibo opens its mouth just as an adult would imitate a baby. This research (funded by the National Science Foundation) is currently investigating whether a robot dog can offer the elderly meaningful companionship over the long term.

GETTING EMOTIONAL

Why mess up a perfectly good robot by giving it emotions? Surely one advantage robots have over humans is that they can be logical and rational and not subject to human emotional weakness? And supposing it had emotions, what sort of damage might an angry or unhappy robot do?

THIS VIEW OF EMOTION AS PRIMITIVE, disruptive, and unworthy of the rational mind is very common. Learning to control our emotional responses is considered an important part of growing up. Yet if emotion had not played a positive role in our evolution, it would never have become such an important part of our makeup.

Emotional intelligence-a term popularized by Daniel Goleman's book of the same name—is a vital part of social intelligence. The Portuguese neurophysiologist Antonio Damasio also suggests that emotion is an essential part of cognition. One of Damasio's patients had suffered brain damage that prevented him from feeling any emotion, but did not impair his ability to think. This patient found it almost impossible to arrive at decisions—he would spend an enormous amount of time and effort weighing all the alternatives without being able to settle on any one of them. This suggests that emotion helps humans avoid some of the huge processing requirements demanded by pure logic, which, as we saw in Chapter 4, makes it so hard to run logic on robots. Emotional intelligence is said to have five components:

Self-awareness: Observing yourself and recognizing a feeling as it occurs;

Managing emotions: Handling feelings in a way that is appropriate; realizing what is behind a feeling; finding ways to handle fear, anxiety, anger, and sadness;

Self-motivation: Channeling emotions in the service of a goal; emotional self-control; delaying gratification and stifling impulses;

Empathy: Sensitivity to others' feelings and concerns, and seeing their point of view; appreciating the differences in how people feel;

Relationships: Managing emotions in others; social competence and social skills.

LEFT

MOOD SWING
The human face is capable of conveying an extraordinarily wide range of feelings and emotional states.

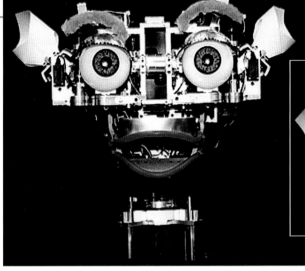

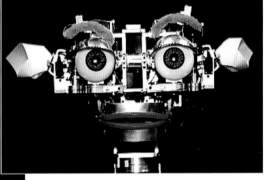

If a robot has no emotions, how can it have emotional intelligence? And if it has no emotional intelligence, how can it interact with humans in a friendly, sociable, and understanding manner?

SHOW SOME EMOTION

Giving a robot an emotional system also means giving it ways of displaying emotion. One method is to include emotion in part of the robot architecture that chooses its behavior. Juan Valesquez, a researcher at MIT, produced a robot that could be scared by a particular blue object and showed its fear by backing away from the object. Even when the researcher hid the object behind his back, the fear reflex would remain for some time, and the robot would hesitate to come anywhere near.

Most emotional expression requires a face, since this is how humans chiefly display and understand emotion. The robot Kismet in the MIT AI Laboratory is the best example of a robot using a face to display emotion. Kismet looks like a robot and not like a person, but it has brown furry eyebrows, soft pink ears, and red squashy lips, as well as big blue doll-like eyes that can convey the sensation of eye contact.

The team that created Kismet drew on the findings of the psychologist Eckman that five or six basic emotional expressions—joy, sadness, fear, anger, disgust, and surprise—are universally recognized by humans across all cultures. Kismet uses these expressions as part of a repertoire of social interactions based on parent-infant behavior. It uses gaze direction, facial expression, body posture, and vocal babbles to signal interest, and also withdraws as a response to overstimulation or crowding. Humans interacting with Kismet report strong feelings of interest and fascination, with many finding themselves responding to the robot as if it were a baby.

Kismet is the result of many years of research effort at MIT Labs. Feelix, on the other hand, created by Dolores Cañamero and Jakob Fredslund at the LEGO-Lab, Aarhus University, demonstrates that a fairly low-tech approach using the LEGO Mindstorms kit can also meet with some success. The ability to express emotion is an essential part of the move toward socially intelligent robots.

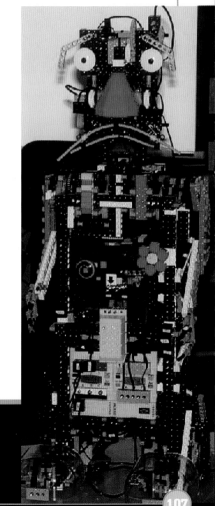

TALKING BACK

We've seen that we can give a robot expressive features to use in social interaction. But how is the robot to know what a human is feeling?

MISERABLE SMILE
Not all smiles imply happiness. According to FACS this smile is a combination of the smile unit AU12, and "sad" action units AU1 and AU4. One way of interpreting this is as a brave struggle against unhappiness.

HAPPY SMILE
This is a smile as defined by Duchenne. In FACS terms it is a combination of AU12, a "lip corner puller," and AU6, a "cheek raiser."

WHEN TWO HUMANS INTERACT, A LARGE number of cues are used and interpreted: words, tone of voice, facial expression, body movement—gestures, the use of eye contact, or looking away.

The mere act of collecting this information is a tough enough task for any robot. Since human-robot interaction can't depend on humans being wired up for the robot's benefit, the robot will have to use cameras, microphones, and other sensors to capture data on the behavior of its human interlocutor.

Current work on computer vision in the field of face recognition for use in automatic surveillance is of some help. These systems employ a general knowledge of the features that make up a face—oval shape, two eyes, one nose, one mouth—to locate the face in the camera field and track the position of the head and its main features. Iain Matthews and Simon Baker, researchers at the Robotics Institute of Carnegie-Mellon University in Pittsburgh, have developed a system that fits a mesh over a face in a video image, lining up specific parts of the mesh with the main facial features.

TAKING TURNS

Once the robot has access to this information, it has to learn what it signifies in an interaction—in a conversation, for example. It would be very rude for the robot to interrupt when the human partner was speaking, so it needs to be able to pick up the signals indicating that the speaker has finished and wants to allow someone else to speak—what is called turn-taking. Humans signal this in several ways—asking a question, sharply raising or lowering voice pitch, using a verbal cue (such as "y'know?"), or a change in eye contact and gesture. Most of us would use several of these cues simultaneously, so the robot has to look for significant combinations.

The robot also has to be able to deal with what is called "back-channel" information while it is speaking. When humans have a conversation, the listener reacts to what the speaker is saying by nodding, making eye contact, adding phrases like "really" or "I see," or repeating some of the phrases the speaker has used. Humans understand that these are responses, not interruptions. If a robot fell politely silent as soon as the human listener said anything, the conversation would never get started.

READING EXPRESSIONS

It's one thing to track facial features with a camera, and quite another to know what the expression on someone's face really means. In the 19th century, a pioneering French neurophysiologist, Guillaume Duchenne du Boulogne, conducted experiments using electric shocks to stimulate particular muscles of the face on a patient inflicted with almost complete facial paralysis. Duchenne photographed the resulting expressions, which he published in *The Mechanisms of Human Facial Expression* in 1862.

In the 20th century, psychologists Ekman and Friesen

produced a system for describing all the facial movements that a human can observe. The system, called the Facial Action Coding System, or FACS, defines action units (AUs) on a face that cause facial movements. Action units are not the same thing as facial muscles, as some muscles produce more than one change in the face: the frontalis muscle, used for eyebrow-raising actions, is separated into two action units, according to whether the inner or outer part of the muscle causes the motion. There are 46 AUs that account for changes in facial expression, and 12 that describe changes in head orientation and gaze.

Jeffrey Cohn, at the Robotics Institute of Carnegie-Mellon University, has developed a system for detecting some of these action units automatically. First the mouth, eyes, brow, cheeks, and their related wrinkles and facial furrows are tracked. Then a neural network, working separately on the upper face and on the lower face, is used to recognize 11 basic lower face action units and combinations and seven basic upper face action units. AU combinations can be recognized as an expression (a smile, for example) and the robot has to try to work out what the expression means—not always easy when you consider all the things a smile can mean.

BELOW

NATURAL SMILE

A genuine smile is one of the most difficult facial expressions to fake. Perhaps this is why it is universally recognized as a reliable indicator of a positive mood.

RIGHT

BABY TALK

A mother is often able to understand her baby's attempts at communication: body language, facial expressions, and cooing—or screaming—can speak volumes.

HOW HUMAN?

It might seem obvious that a robot designed to interact with humans should look as human (or anthropomorphic) as possible. But are we sure about this? It isn't easy to make a robot look really human, so it would be reassuring to know that it was worth all the effort.

THE UNCANNY VALLEY

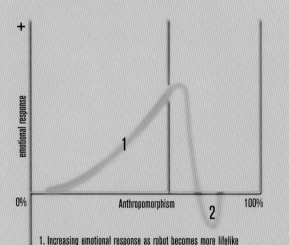

1. Increasing emotional response as robot becomes more lifelike
2. Reversal of positive feeling—the uncanny valley

Masahiro Mori considered the relationship between how familiar—or humanlike—a robot seemed and an individual's emotional response to it. He concluded that there was a sharp drop in positive feeling towards the robot at the point at which it began to look nearly-but-not-quite-human, and called this the "uncanny valley."

ALLISON BRUCE, A DOCTORAL STUDENT AT CARNEGIE-MELLON University, carried out an experiment using a robot with a laptop computer mounted on the top. This would stop passers by at random in the University corridors and, in a synthesized voice, ask them to answer a short questionnaire. Sometimes the laptop would have a human face filling its screen, and sometimes not. The results showed that people were more willing to stop and be questioned when the robot appeared to have a face.

If a face is useful, does it have to look very like a human? Hiroshi Kobayashi's work, at the University of Tokyo, has concentrated on Face Robots—robots with the mechanical equivalents of human facial muscles covered by a latex skin, a combination that can create accurate human expressions. We have seen that the MIT robot Kismet can involve humans in social interaction without looking very human at all.

The more real a humanoid robot looks, the more human its behavior is expected to be, and this can create unforeseen problems. We all accept that cartoon characters do not do lip synchronization—the animator just opens and shuts the cartoon mouth more or less in time with the accompanying speech. But viewers watching a dubbed film may be irritated by the mismatch between the lip movements of the actors and the sound of the words. The difference is that we expect perfect lip synchronization from human actors.

Encountering a very human-looking robot, we may assume that it can understand a joke, be aware of nuances of expression, or make small talk. If the robot fails to live up to these expectations, we are likely to become irritated and critical.

Sony Aibos are clearly made of metal and plastic. They do not have furry suits; they are robots and not real dogs. But their behavior has been cleverly engineered to be doglike. Humans seem to find them attractive and interesting interaction partners—suggesting that a robot's patterns of behavior are far more important than its appearance.

THE UNCANNY VALLEY

In the early 1980s, Japanese researcher Masahiro Mori tried to predict the psychological effects of increasing degrees of humanness in robots and dolls. His key prediction was that, as a robot or doll becomes more similar to a human, the human response to it gets more positive, right up to the point where it appears very nearly human. But a sharp drop in sentiment takes place at the point where the robot might almost be mistaken for human when some small discrepancy suddenly reveals that it is not, producing a psychological shock. He called this the "uncanny valley," but it might also be called the zombie effect.

A robot can do many things that a human cannot. But giving it a very human appearance may end up having the opposite effect to that intended, making people see it as less than human. This may disappoint or unnerve them, and make it harder for them to understand and use the robot's special non-human capabilities.

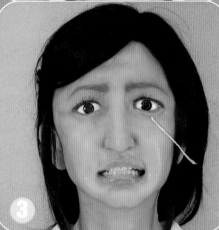

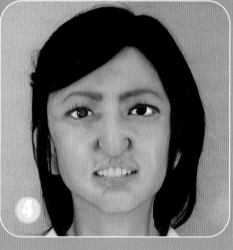

ABOVE & RIGHT

FACE MASKS

Hiroshi Kobayashi's Face Robots use plastic
and other soft materials over a metal
infrastructure that acts like the human facial
muscles. This allows the robot to reproduce
standard human expressions—here we see
(1) happiness; (2) anger; (3) sadness;
(4) disgust; and (5) fear. But is there a danger
that these robot faces may evoke Mori's
"uncanny valley" response in a human partner?"

PLAY WITH ME

Roboticists have been trying to get their creations out into the real world for many years. Perhaps they hadn't expected that this would first happen through robot toys—yet toys have always been ideal candidates for robot capabilities.

A TOY THAT MOVES ON ITS OWN HAS ALWAYS BEEN A great attraction, from train sets and clockwork toys to radio-controlled cars, boats, and planes. Dolls have been made with small speech components (operated by a string or button) and the ability to cry, drink, and wet themselves. Children are perhaps more open to new ideas and novel abilities than many adults, while the high-tech tag is a selling point for their parents. And a toy doesn't have to perform a specific task usefully or accurately—it only has to be interesting and fun.

The Sony Aibo robot dog was originally presented as a toy for adults—a robot pet for families that didn't have the space or the energy for a real dog. But the ground had already been prepared by the small devices that took Japan by storm in the mid-1990s: Tamagotchis, which engaged the caring instinct of so many children (and not a few adults).

The Tamagotchi was originally the size of a digital watch and had a tiny LED screen on which some very low-resolution graphics indicated its state and feelings. It needed to be played with to keep it happy; be fed when hungry; injected when sick; have its light turned off at bedtime; be scolded when naughty; and even cleaned up after when it used the toilet. It could not be left unattended for a day or it would signal that it was hungry and unhappy. If neglected, it would die.

The success of the Aibo has led to a new generation of Tamagotchis. These look like robot cats, dogs, or babies, have some limited ability to move around, and in some cases touch or infrared sensors to support more sophisticated interaction than was possible with the tiny buttons that had to be pressed on the originals. They are far less expensive and less sophisticated than the Aibo, but still a big step toward mass production of robot toys.

BABY DOLL

Dolls, with the expectation of human behavior already attached to them, are an obvious choice for a robot toy. In the late 1990s, MIT spin-off corporation iRobot and toymakers Hasbro adapted the technology of the MIT AI Laboratory to create "My Real Baby." This robot doll has facial expressions—smiles and frowns—as well as reactions to physical play,

ABOVE

WHAT COULD BE MORE FUN?
Maybe robot dolls have more to do with adults being carried away by technology than with the real needs of children.

ABOVE (INSET)

PETTING AIBO
Peter Kahn and Batya Friedman of the University of Washington, and Alan Beck and Gail Melson of Purdue University took a Sony Aibo to school to see how children responded to it, in a project partly funded by the U.S. National Science Foundation.

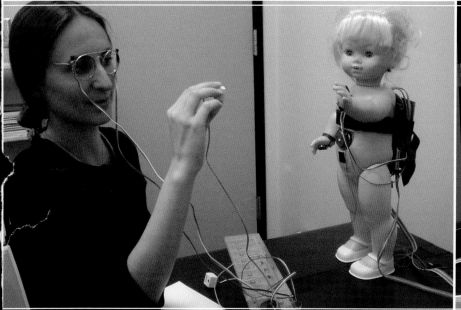

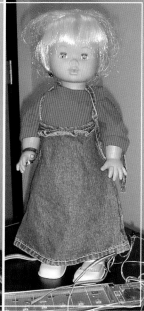

BABY BOT

Robota the robot doll was designed for experiments in human-robot interaction rather than as a commercial product. She can dance and imitate the behavior of a human interacting with her.

such as giggling when you tickle her feet. She also has some learning and adaptive abilities, so that her behavior changes and becomes more independent over time.

The Robota doll was originally a research prototype by Aude Billard, a doctoral student at the University of Edinburgh in Scotland. She wanted to investigate how giving a doll the ability to imitate a child's behavior would affect interaction, as well as to look into the effects of a range of interaction abilities, like reacting to touch and handling (inclination), to human presence (pyroelectric sensor and camera), to simple joystick commands, to music (CDs and xylophone-keyboard), and to speech.

ABOVE LEFT

FOLLOW MUMMY

Robota's creator, Aude Billard, wears tracking hardware on her glasses and hand so Robota knows where she is looking, where her hand is, and can imitate her grasping action.

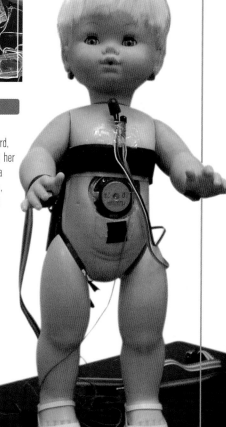

A GOOD IDEA?

Many experts in child development are unhappy about robot dolls. The move to realism seems to them to cut off open-ended imaginative play in which the child supplies the speech and behavior of the doll. They also have concerns that pre-school children who are still learning to distinguish between reality and fantasy may be confused about whether the doll is really alive, and that this realism will discourage children from playing with other children. Would funding playgroups be a better use of resources than developing these expensive robot toys?

HELPING AND GUIDING

Few people would be surprised to see a robot in a museum of science and technology—unless, of course, the robot turned out to be their tour guide.

RHINO, THE FIRST EVER ROBOT TOUR guide, made his debut at the Deutsches Museum in Bonn, Germany, in 1997. For six days he guided hundreds of visitors around the museum from one exhibit to the next. Rhino's development team involved researchers from the University of Bonn and Carnegie-Mellon University, and the latter drew on this experience in 1998 to produce Minerva, a robot guide for the Smithsonian Institute in Washington. Minerva's job was to guide visitors through the "Material World" exhibition of artifacts showing how materials have influenced the way humans live, and it used a recorded human voice to explain how robots fitted into the history of the items.

Minerva was a standard wheeled robot, humanized by a movable red mouth and blue eyebrows that allowed it to smile or frown. It would welcome approaching visitors by smiling, and used the frown when people crowded it, intoning "Excuse me, I need to get through." It also had a horn to clear a path. The use of voice and expressive face seemed effective in getting people to treat Minerva with an almost human degree of respect—though at one point some kids jumped on and tried to take the robot for a ride, illustrating the point that safety issues apply both ways.

While you might conclude that a human guide would be far better than a robot for visitors who are physically present, a robot guide can be linked to the Internet and allow someone on the other side of the world to share its physical tour. Now researchers in Greece and Germany, funded by the European Union, are collaborating to make a "TourBot" that can be used just like this.

ROBOT NURSE

Museum touring involves human contact, but the robot does not have to build a lasting relationship with the people it guides. Creating a domestic robot to help the elderly remain independent at home is a more challenging project. The social and emotional issues of interacting with one elderly person and their carergivers—who may have no enthusiasm for new technology—are far more difficult. A group at Carnegie-Mellon University and the University of Pittsburgh is working on a "Nursebot," which they hope will be able to manipulate appliances for elderly sufferers from arthritis. A team at the Fraunhofer Institute in Germany is working on a "CareBot" with similar aims.

The U.S. group is also investigating what it calls cognitive orthotics—the idea of using a robot to help with cognitive problems, such as unreliable memory. The robot would remind its client about mealtimes or medication intervals and—because it is able to observe the residents—should know whether the required actions are actually performed. If, for example, three doses of medicine are missed in a row, the robot could inform the carergiver. Clearly there are sensitive human issues to negotiate here, and in the circumstances equipping the robot with social knowledge and skills will be at least as important as getting the underlying technology to function.

ROBOT THERAPY

Kerstin Dautenhahn of the University of Hertfordshire in the U.K. is one of a number of researchers investigating whether robots can help children suffering from autism. Autism impairs interaction and social skills; autistic people tend to concentrate on detail and literal meanings, and are not good at seeing the whole picture. They can easily be overwhelmed by the number of stimuli and sheer variety of interactions with other people, and the frustration and stress can sometimes lead to behavioral problems. The much simpler and more predictable interaction style of the robot may actually be an advantage. The researchers are exploring whether interaction with robots can gradually take an autistic child from simple interactions through to more complex and varied ones, one step at a time.

One major question is whether robots used for this purpose should be simple and machinelike in appearance, or whether a more human exterior is useful. Robota, the robot doll introduced on page 113, is now taking part in the experiments.

CREATIVE SPIRIT

We think we can use robots to help and guide humans, and maybe even for therapeutic purposes. Now even more creative applications are being investigated.

THE "PETS" PROJECT AT THE UNIVERSITY OF MARYLAND IS much more than just another robot animal tale—together with a group of local children, the Maryland team has designed a Personal Electronic Teller of Stories. The children were full partners in the design of these robotic pets that help them tell stories about how they feel. The kids could build any animal they wanted using a set of robotic animal parts (dog paws, wings, horns, and more). The team has produced "My PETS" software to support story-telling, and this includes giving the robots emotions and behaviors.

The first prototypes were produced after two full weeks of effort in the summer of 1998. There were three design teams, each involving two adults and two children: the skeleton group, the skins and sensors group, and the software group. The groups had to interact—skins had to fit skeletons, and both had to be strong enough. The software group picked out seven feelings they wanted the robots to be able to display, listed the actions that would show each feeling, and then tested to see if people could guess the feelings when team members acted as the robot would.

ACTING UP

The storytelling theme is being taken up by other groups of researchers who are trying to produce robot actors. This may seem an odd idea, but there are already several dramatic traditions that involve the manipulation of large puppets—Japanese Bunraku puppets, for example—and robots can be thought of as self-animating puppets.

ABOVE

ANIMAL TALES
Two University of Maryland PETS robots, bodies and base.

RIGHT

STORYBOT
A PETS robot complete with the software used to write stories for it to perform.

Barry Brian Werger of the University of Southern California ran an early project with Ullanta Performance Robotics in the period 1995-98. He argued that robot actors were both a novel artistic idea, and an excellent testbed for robotic technology—a robotic performance piece had to appeal to audiences, take place at scheduled times, and meet a very high standard of robustness, adaptivity, and interaction.

The most popular Ullanta piece was called "Fifi and Josie, A Tale of Two Lesbiots: A Story of Autonomy, Love, Paranoia, and Agency," a traditional scripted play written by Roxanne Linnea Rae and performed in Boston in 1995 and 1997, and in Scotland in 1997.

A small project at Carnegie-Mellon University in 1998-99 aimed to take this a step further with the idea of Robot Improv. "Improv"—improvization—is a form of acting in which actors are given basic character and situation, and usually an overall goal, but no script. The Robot Improv robot actors performed a classic Improv exercise for two actors, in which one actor tries to leave the room while the other uses a variety of methods to try to stop him or her, such as becoming violent or pleading. The project team created a number of robot characters with different qualities— unfriendly; friendly and trusting; friendly but manipulative.

WHOSE FACE?

Robot actors may not be everyone's dream. Two actors from the sitcom *Cheers* sued when Paramount Studios licensed a company to run *Cheers*-style bars in airports across America. The bars were to be equipped with a pair of robots based on the actors' *Cheers* characters, and when the actors complained about this, the company just changed the robots' names. The actors were clearly not amused.

MARCEL MARCEAU
Could a robot ever achieve a performance to rival that of the great mime artist?

ME AND MY PET
P. Kahn and B. Friedman, University of Washington, and A. Beck and G. Melson, Purdue University, found that robots can stimulate all kinds of creative play in the hands of willing volunteers.

Keeping Up With The Competition

EVERYONE LOVES A COMPETITION—BUT ROBOT competitions are not just for fun, but to help push research forward.

The AAAI (American Association for Artificial Intelligence) has been running its annual competition since 1992—the oldest AI-centric robot competition in the world. Although probably inspired by human athletic contests, the AAAI does not run exactly the same event each year, but invents new and more difficult challenges as robot capabilities improve.

Back in 1992, the competition arena contained easily detected obstacles along with marked objects that the robots had to search out in a series of tasks of graded difficulty. By 1995, all the robot entrants were able to complete all the tasks, so in 1996 robots had to collect tennis balls—including some battery-powered moving balls—scattered all over the arena and place them in a bin.

DUMB WAITER
Joint second in the 2001 "Hors D'oeuvres, Anyone?" event, the Seattle Robot Club entry was built completely from inexpensive, off-the-shelf hardware.

In the "Hors D'oeuvres, Anyone?" event, added in 1999 and still running today, robots are required to serve food to the conference's human delegates as they wander around the robot exhibition that accompanies the competition.

For the first time, the robots had to operate in a human space rather than a special arena. This obviously required them to be able to avoid running over the delegates as they did their rounds. Covering the area and finding groups of people to offer food to was a second necessary ability, and a third was noticing when the food tray needed refilling. In the 2001 version of this event, the robots also had to wear a server's badge and interact with other robot servers.

WINNING WAYS

The 2001 winner, José from the University of British Columbia, Canada, combined many features we saw in earlier chapters.

It used three calibrated video cameras to give it stereo vision, and a webcam that monitored its black tray of hors d'oeuvres, deciding the tray was empty when the proportion of non-black pixels got too low. An extra pan-tilt camera enabled it to find people to serve, by looking for clusters of skin-colored pixels. It worked out its position in the room using odometry. With its stereo camera cluster, it collected a database of visual landmarks to use in correcting its position estimate.

An animated face on the laptop screen on top of the robot and a pleasant tenor voice to offer food and make a few simple jokes completed the winning robot server. As a final touch, it would scold people who took food before it was offered, and at least one delegate was observed guiltily putting food back and apologizing.

URBAN SEARCH AND RESCUE

In 2000, a new Urban Search and Rescue event was created. The arena for this event was designed with NIST (the U.S. National Institute of Standards and Technology) and has yellow, orange, and red sections of increasing difficulty. The main objective is to identify victims and send their locations back to the medical teams, but minimizing the number of robot operators is also important. The scoring system also takes into account the percentage of victims found, the number of robots finding each victim, and the accuracy of the locations reported.

Entrants in the summer 2001 event were not to know that some of these robots would be deployed for real in the wreckage of the World Trade Center just a month later. In real urban search and rescue, as experience in New York showed, intelligence and autonomy are not enough on their own. A wonderfully intelligent robot that lacked the mobility to get through a terrain of crushed buildings would be no use after an earthquake or terrorist attack. In the 2001 competition, only one robot had the mobility to operate in the red zone.

Real-world rescue workers do not want to be kept out of the decision loop, so teleoperation—driving the robot remotely—is a favorite option. The ideal robot would be semi-autonomous, so that a human could direct it strategically while it dealt with problems on the ground.

A great aid to real-world use is the umbilical—a cable supplying power to the robot and carrying back the video signal from its cameras. This solves the problem of a robot running out of battery power, and enables it to be located and maybe even rescued if it breaks down in the middle of a wrecked building. The umbilical also makes direct communication possible where radio signals are blocked by twisted heaps of metal and concrete.

The AAAI competition aims to encourage not only world-class robotics engineering but the development of a robot equivalent of smartness. This is robotics not just for the research lab but for every-day interaction in the human environment.

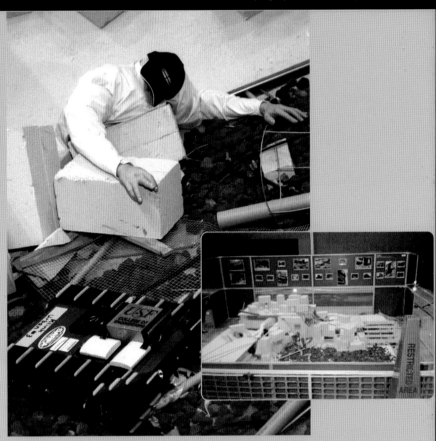

[↑] THE RED ZONE
This area of the Urban Search and Rescue event was designed to mimic a flattened building. Much worse was confronted for real the very next month in the collapsed World Trade Center buildings.

[↗] FLORIDA RED
A University of South Florida team, under the leadership of Robin Murphy, has been working on the Urban Search and Rescue problem for years. Here, its URBAN robot traverses the red zone.

HOBBYBOTS

Finally, let's confront our age-old fear: are robots destined to replace humans as the dominant species on Earth? So far we've examined just about every aspect of robot development—movement and sensing, energy, learning, language, and interaction with people. We've seen that, though roboticists are trying hard to learn from living things, robots still lag behind biological creatures in both structure and capabilities. They remain machines—hampered by limitations and a long way from being "living robots."

The simple answer to the question above is: "Not until they can last more than two hours without recharging their batteries, and can climb stairs as fast and efficiently as a human." But let's take a more in-depth look at the obstacles. What would robots need to gain autonomy as a species, autonomy in the sense that animals know it? Unless they do, it's hard to imagine how "taking over the world" could even begin to make sense.

WHAT KIND OF WORLD?

If we could create a robot species, what kind of world would it inhabit? Let's say we were to produce a self-sustaining species of cleaning robot: small, specialized creatures that scurried around the insides of buildings, vacuuming and tidying. Like bees, who benefit humans by pollinating plants and making honey, they would live their own autonomous lives in parallel with ours. We might create not just one of these robot species but dozens, each with its own purpose. Nanorobots might inhabit the human body just as bacteria do, clearing cholesterol out of blood vessels.

Chapter 7

SO WILL THEY TAKE OVER THE WORLD?

But this type of living robot would be designed by humans to fill a particular niche in society and stick to it. These "robots of the world" would be no more likely to unite than the planet's reptiles would be to gang up against the human race. When people worry about world domination by robots, they picture a society controlled by humanoid robots. But such a scenario is even less likely—a robot species with human characteristics would have to have everything that specialized robots require and a great deal more. It would need not only humanlike abilities at the individual level, but also undreamed-of social skills—whether trying to live in the human world or construct one of its own.

ONLY CONNECT

Increasing globalization has made us aware just how interconnected our world is. Even the ingredients for a simple meal could come from a dozen different countries. Hundreds of people will have been involved: in farming, manufacturing, processing, transportation, and logistics.

We take our high level of social organization for granted unless and until it is disrupted by war or natural disaster. To rival us, robots would need the ability to organize a self-sustaining social fabric of similar complexity. But at this stage, when giving robots individual abilities still remains problematic, the question of a robot society has barely begun to be researched. We will see, however, that experimental robot ecologies already exist.

BACKGROUND

CAR PLANT
Industrial robots have little or no sensing capability, and are useful because they can repeat a very simple, highly structured task indefinitely.

RIGHT

EYEBALLING
Professor Noel Sharkey, of Sheffield University, United Kingdom, looks his predator robot in the eye. The predator is one of a set being built for a robot ecology at the Magna Science Adventure Centre, in Rotherham in the United Kingdom.

WHAT'S IT ALL ABOUT?

What makes living things autonomous? We know that one answer is autopoiesis—the way in which the interaction of living things with the environment maintains the internal processes that keep them alive. This includes the ability to grow, to heal wounds, and to fight off bacteria using immune systems, as well as breathing, digestion, excretion, and circulation. And we have seen that the standard robot, made of metal, batteries, and a computer, does none of these things.

SHARING CULTURE
Humans can pass on information that cannot be carried in genes. Culture allows each generation to use the experience of previous ones, not just genetic material.

ANIMALS HAVE TO BE ABLE TO MOVE around their environments without damaging themselves by crashing into rocks, falling into holes, or toppling off cliffs or into rivers. They have to find food to maintain their energy levels. They must avoid hazards, from bad weather to other animals out to eat them. What would it mean for a robot to be self-sufficient in this kind of way?

For animals that reproduce sexually, the point of all this effort is to find a mate and produce viable offspring. That way, the species can continue as long as it can adapt to major changes in its environment, whether caused by the climate or the presence of other animals. A species that reproduces and adapts successfully does so because it occupies a stable niche in the overall ecology. What is the robot equivalent of sexual reproduction? How can robot ecologies be brought into being? Work in these areas is really only just beginning.

A ROBOT CULTURE?

A major difference between humans and other animals is that—over and above our sense of ourselves as individuals or local groups—we are aware that we belong to a species. And we have language, which creates a powerful new mechanism to transmit information between generations. This means that humans can pass on knowledge from one generation to the next, instead of each generation having to learn everything for itself. So the experience and the understanding of the whole human species grows all the time. For most animals, transmission is carried out almost entirely by passing on genes to their offspring.

If living robots were to have a culture, they would first have to develop a language in which to express it. The "Talking Heads" experiment described in Chapter 5 is an early attempt to have robots construct a language that corresponds to their own perception of experience, instead of trying to program them to speak human languages. The structure of such a language would depend heavily on why the robot needed to communicate, and what about. It might prove to be very simple, more like communication between say dogs or cats than humans.

Language is only the first step in creating a culture. Robots would also have to develop the social mechanisms to sustain it—social gatherings, cultural artifacts, arts, sciences, and the creative communities that support and develop them. They would have to develop mechanisms for passing culture between robots—the equivalent of our education.

Because we use speech or writing to communicate, while robots could use radio transmission or connect directly to the Internet, they might seem to have some advantage. But culture is selective—not every experience or every action of any individual becomes part of it. Robots would have to learn this process too.

It's hard to imagine how we could give robots a species consciousness when they do not yet have an individual consciousness. And it's even harder to imagine giving robots an individual consciousness when we still have no idea what produces consciousness in humans. So robot culture is another concept whose time has yet to come.

BACKGROUND

JUST CHATTING?
Language is the basis of culture. But what we talk about is our common experience—what would robots have to say to each other?

BELOW

GO FORTH AND MULTIPLY
Finding a mate and producing viable offspring is what makes a species live on. What is the robot equivalent?

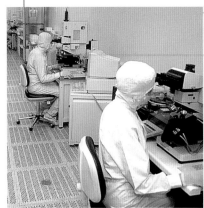

ABOVE

ROBOT OBSTETRICS

A new robot depends on the design and production of silicon chips.

RIGHT

TOUGH ON THE OUTSIDE

Robot reproduction really starts with digging out the metal ores that will eventually make up the robot body.

REPRODUCTION

If robots were to become a species, they would have to be able to reproduce—that's really what a species is, an organization of living things that can maintain itself beyond the life of individual members. What would be needed for robots to be able to reproduce themselves?

FOR ANIMALS, REPRODUCTION IS A VERY local process. Sexual reproduction needs two animals to mate in order to combine their genetic information. Building the new individual uses the resources of the local environment, whether the combined genetic material is left to its own devices, or kept in a protected environment, such as a nest or inside the body of one of the mates.

The process of making a new robot is rather different. Ores have to be mined and smelted into metal or metal alloys, then molded or machined into components. Plastics have to be constructed from hydrocarbons; silicon chips must be prepared and their circuits laid down using ultra-violet light and optical miniaturization facilities. Electrical cables are produced, and motors assembled.

The instructions controlling the production of a new animal are also local in the extreme—embedded in the animal itself, in the genetic material it acquired from its parents. A robot design, by contrast, is not stored within the robot: in the absence of detailed plans, the only way to find out how a robot is put together is to dismantle it.

The way genetic information expresses the design is also unique to living things. In a design for a robot, we would expect to see an image of the finished robot together with diagrams indicating how the components should be assembled. The relationship between the genetic information held in an animal—the genotype—and what a grown animal looks like—the phenotype—is much less obvious. The genotype doesn't hold a picture of the phenotype; instead it contains the control instructions for the processes that will construct the animal. Growth is very different from assembly.

MIXING IT UP

Sexual reproduction in animals ensures that no two individuals are identical. The aim of robot design is the opposite: each model should be exactly like all the others. In practice, no two robots will be identical—the wheels may slip a little more or less, the dimensions may vary slightly—but these are differences of performance, not design.

Because in a living species each individual is unique, a change to the environment that disadvantages some individuals may benefit others. It is this genetic variability that enables a species to adapt to the environment over long periods of time. (It is, incidentally, the lack of genetic variability that makes cloning such a risky strategy—a new disease that kills one clone will also kill all the others.)

Human genetic variability has given us the ability to survive repeated attacks by new and deadly diseases—think of the Black Death in 14th century Europe—because some individuals have had enough resistance to overcome them. Because robots have not yet been called on to survive for long periods—tens of thousands of years, say—as a species, any problems inherent in the fact that all robots of a given design are essentially the same have not yet showed up.

SELF CONTROL

Animals can reproduce without a complex society to back them up, just as long as they can remain self-sufficient for long enough. But producing even the simplest of robots needs a huge and complex social organization. It's hard to see how robots could replace the human infrastructure they depend on with a robot infrastructure that would give them control over their own reproduction. And without control over their own reproduction, how could robots ever become a species?

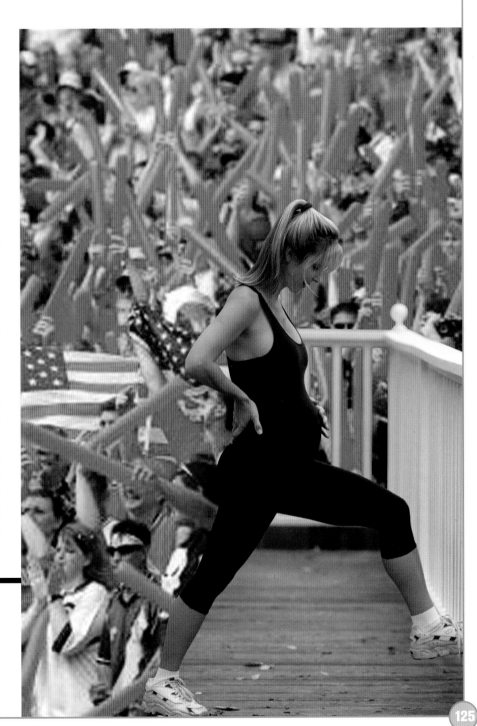

TOP RIGHT

VARIETY—THE SPICE OF LIFE
Sexual reproduction ensures that individuals are different, giving the species many options if the environment changes.

FAR RIGHT

PROTECTION
Human babies are constructed locally, inside the body of their mother.

THE STATE OF THE ART

A true story: a researcher attempts to board an aircraft while carrying a suitcase containing Max, the six-legged robot we met in Chapter 4.

SUSPECTING THAT THE OPERATOR ON THE BAGGAGE SCREENING machine might be startled to see all those motors and wires (though fortunately this was before recent concerns over terrorist attacks) the researcher announces Max in advance to the security staff. They insist sternly that the suitcase be opened and Max be displayed. The researcher proudly lifts Max out of the suitcase and sets him down in front of the assembled security workers. "Is that it?" says one, clearly very disappointed. "We were expecting something like Star Wars."

Many roboticists must wish Star Wars had never been made—measuring the current state of the art in robots against science fiction always seems to disappoint expectations. It's hard to find a balance between fears of robot domination on the one hand and "Is that it?" on the other. So what is the current state of robots in real-world applications?

The sensing technologies that have made their way from robot research into real-world applications are often barely thought of as robot-related, at least not in the Star Wars sense. Cruise missiles are effectively robots, using machine vision as well as Global Positioning System technology to navigate accurately over terrain to a selected target.

In other cases, the barriers separating robots from the real world are related to cost rather than technology. Domestic cleaning robots are a success story waiting to happen— but cleaning can be done easily and cheaply by human effort.

Most cleaning robots use laser scanners, ultrasound, infrared, or a combination of these technologies to navigate, but as they only have to cover the ground while avoiding obstacles, their sensing requirements are not demanding. Autonomous grasscutters use the same technology—but the issues of safety and legal liability in a machine with sharp blades attached to it form another barrier to widespread use.

ROBOT CARS

Autonomous cars are already a reality. In a European research project in the 1990s, Daimler-Benz—now DaimlerChrysler—developed a vehicle that drove autonomously for 1864 miles (3000 km) on highways in Germany and France. The driver remained in the front seat, and could bring the car back under full human control by turning the steering wheel or depressing a pedal. By 1998, a version of this device was brought to the mass market: the Autonomous Intelligent Cruise Control.

Regular cruise control allows the driver to set a speed that is maintained autonomously —useful on long, relatively empty highways, as in the United States or Australia, but more problematic on crowded roads, as often in Europe. DaimlerChrysler added a radar sensor to determine the distance from, and the relative speed compared with, the car in front. Should the distance dip below a set limit, the system reduces engine output and activates the brakes.

Widespread use of this technology could see cars forming into convoys on busy highways, probably reducing traffic jams and accidents. But although DaimlerChrysler's research vehicle could also perform autonomous overtaking maneuvers, this feature has not yet made it into production. Again, legal considerations would seem to constitute the main barrier.

HELP ON WHEELS

The HelpMate Trackless Robotic Courier can be seen in some United States hospitals, delivering and retrieving trays, linens, routine medication, records, and other items. It is equipped with radio communication, so that staff can summon it and it can summon elevators. It also has a synthesized voice—male or female on request—with some pre-scripted phrases.

It's no accident that these real-world applications of robot technology nearly all involve wheeled vehicles—legged mobility is still an immature skill for robots. Complex sensing, such as vision systems, is also still too unreliable for most real-world applications. But robots are gradually edging out of the labs and into the real world.

LEFT

CLEANBOT
Cleaning robots are now technically feasible—but still too expensive for widespread use.

ABOVE

END OF THE ROAD RAGE?
Intelligent cruise control would allow cars on a highway to act like robots and form a convoy, reducing accidents and jams—and driver frustration.

RIGHT

STAR WARS
Expectations raised by films such as *Star Wars* are hard for real robots to meet.

ROBOT ECOSYSTEMS

Despite our technological prowess, we humans have come to realize that we depend completely on the Earth's ecosystem. Concerns over global warming and genetically modified plants reflect fears that we could destroy a balance we don't fully understand.

NOSEY

The predator robot in the Magna Science Adventure Centre in the UK–seen here with designer Professor Noel Sharkey–has a long snout, allowing it to extract power from its robot prey.

AN ECOSYSTEM CAN BE DEFINED AS A DYNAMIC COMPLEX OF plant, animal, and microorganism communities and their non-living environment, interacting as a functional unit. Can we imagine an ecosystem that includes robots as a species?

Robot self-sufficiency would be a first step, and the idea of a robot ecology goes right back to the late 1940s with William Grey Walter and his tortoise robots, Elsie and Elmer. As we saw in Chapter 3, Grey Walter's robots could independently enter a hutch to recharge—but this ecology contained just two similar robots. An ecosystem worth the name must have more than one type of creature.

In a real-world ecosystem, animals interact through predator-prey relationships, otherwise known as the food chain. At its simplest, one animal eats vegetation, and is in turn eaten by another. If the predator eats too many of its prey, or has too many offspring, the numbers of prey drop below the level needed to sustain the predator. If the predators die out, the numbers of prey increase to the point at which they run out of vegetation.

ROBOTS AND PARASITES

Since robot reproduction is still some way off, experiments in robot ecosystems don't involve battle-bot style predators smashing up robot prey. But researchers have tried to build in the sort of conflict relationship seen in nature.

In the 1990s, the AI Lab of the Vrije Universiteit Brussel, in Belgium, worked with David McFarland, a distinguished animal ethologist—an expert in animal behavior—from the United Kingdom, to construct a robot ecosystem. Robots can be given the behavioral trait of trying to recharge their batteries when their power runs out. The Belgian team's idea was to make the power source a resource that would be competed for, and to create an equivalent of parasites who would also consume power.

The robots were made from LEGO, with electronics and computer facilities added by the lab. The charging station was illuminated by a light, as in Grey Walter's system, so that a robot with light-seeking behavior would be able to find it. Attached to the charging station were the parasites: canisters emitting polarized light, and consuming power at the same time. The robots were given another behavior that attracted them to polarized light, and when this led them to crash into a canister, its ability to draw power from the charging station was temporarily halted. The robots did no planning or reasoning, but were wholly driven by their behaviors for light seeking and power recharging. No cooperation was programed in, but if one of the robots knocked out the parasites, another robot would be able to recharge. This led to a simple, self-sustaining ecosystem.

PREDATORS AND PREY

In the Magna Science Adventure Centre at Rotherham in the United Kingdom, a predator-prey robot ecosystem is being set up. Two types of robot have been designed—prey robots who find their "food" using light sensors and predator robots with long snouts that allow them to connect to prey robots and extract power from them. The predators have to survive by stalking and chasing the prey robots for their power.

The experiment should be able to sustain itself for long periods, and will be conducted in public under the eyes of visitors to the museum. If successful, it will be a powerful demonstration of an artificial ecosystem in action.

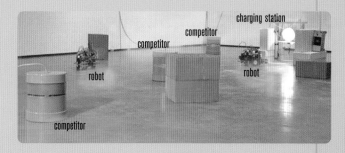

ABOVE

ECOSYSTEM
The VUB ecosystem involves robots and parasites that compete for the power in the charging station. By knocking out the parasites, the robots are able to recharge their batteries.

LEFT

EAT AND BE EATEN
In a natural ecosystem, predator, prey, and vegetation have to stay in balance for sustainability. It takes a large population of antelope to support only a few big cats.

MAKING A LIVING

How would a robot fare in the following simple test of autonomy? Could it survive over an extended period—weeks, months, ideally years—without human intervention?

THE ROBOT'S FIRST CHALLENGE WOULD BE identical to the one faced by all living animals—finding enough "food" to keep going. Until the new energy sources discussed in Chapter 5 mature, this means finding a way of recharging the relatively short-lived batteries that power nearly all robots.

As anyone who has run robot experiments for any length of time knows all too well, running out of battery power is just one of the many annoying mishaps that can occur. Gears can jam, motors can seize, and then burn out altogether. Electrical connections start to come apart, and if not repaired will fail— first intermittently, then permanently. A self-sufficient robot would have to find a way of maintaining itself in the face of such problems.

As we have already seen in Chapter 5, researchers have been producing robots that can recognize and head for recharging points when their battery power is low. This can be achieved relatively easily indoors, but an outdoor robot could not rely on finding a power point. Planetary rovers use solar power, but solar cells don't provide a rapid enough recharge, and in Earth's cloudy atmosphere they are only consistently reliable in climatically favored locations.

A team at the University of the West of England has made an imaginative attempt to combine outdoor self-sufficiency with a useful task in a robot ecological niche.

Vegetation provides too little energy to power a robot, so they reasoned that theirs should carry out the ecological role of a predator. They then looked around for a living thing that would make appropriate prey without raising tricky ethical issues. They found the ideal candidate in the form of the common slug. These are both an expensive pest for farmers, especially in fields of crops such as brassicas, and provide potential concentrations of energy for a robot predator.

MAKING A SLUGBOT

An animal that lived on slugs would extract energy through its digestive system. The designers of the Slugbot decided that this was going to be much too difficult to replicate. But what if the slugs were put into a biomass generator? The slugs would be fermented to produce methane, and this could be used to generate electricity, which would in turn recharge the robot. The loop would be closed by having the robot capture the slugs and drop them into the biomass generator.

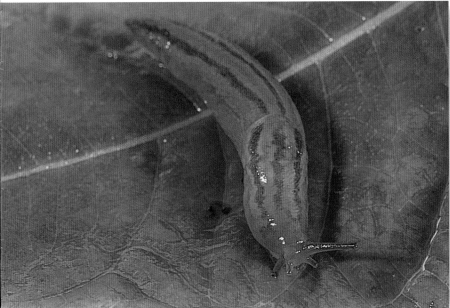

The first challenge was to locate the slugs—trickier than it sounds, especially as they are nocturnal. The team decided to equip the robot with an electronic flash camera programed to take a photograph of the ground every couple of minutes. Anything of about the right size that had moved between shots would probably be a slug. Some vision processing was added to try to exclude the possibility that the robot would attempt to capture slug-sized stones.

HOW MANY FINGERS?

Once the slug is found, the next trick is to catch it. The Slugbot was equipped with large wheels and a long, thin arm that could extend over a wide area, to minimize the distance it had to travel over the soft terrain.

The first design for the hand had two fingers, one to go on either side of the slug. But the team discovered that as soon as the fingers began to grip, the pressure would squeeze the slug's slippery body from between them. So they designed a new hand with three fingers that could simply scoop them up.

Having managed to pick a slug up, putting it down into the biomass generator also proved to be a problem. The slug's slime caused it to stick to the robot's fingers. In the end they had to add a mechanism like a windshield wiper to scrape the slug off the fingers when the hand opened.

The lesson here is that some amazing engineering is needed to produce the sort of adaptation that evolution produces in real animals. Gearing the robot to be self-sufficient in a cabbage field is a challenge the Slugbot team is still working on.

TOP LEFT
PRACTICALLY GENIUS
The Slugbot had to be equipped with three fingers to catch the slugs in a scooping movement, and a scraping mechanism to stop them from sticking to the fingers.

TOP CENTER
EXTERMINATOR
Slugs are a major pest of brassicas and can munch their way through a large crop in next to no time. A robot exterminator would be a dream solution for farmers.

TOP RIGHT
DISTINCTLY SLUGGISH
The Slugbot was designed with big, squashy tires for muddy fields and a long, thin arm to reach out around it.

SELF-DESIGN

Producing a robot is, as we have seen, an extremely laborious process and depends heavily on human effort. How far could this be automated and then controlled by robots themselves? In Chapter 4, we saw how the mechanisms thought to drive evolutionary change in living creatures can be programed on a computer and then used to evolve virtual creatures. Could this idea be employed to allow autonomous design of robots?

SELF-BUILD

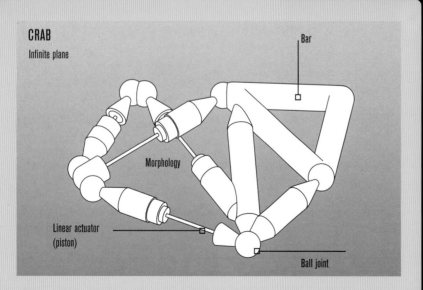

CRAB
Infinite plane

Bar

Morphology

Linear actuator
(piston)

Ball joint

In the Golem Project, the robots are built up by the addition, modification, and removal of building blocks—in the form of bars and joints—over hundreds of generations. The effectiveness of each body shape, or morphology, is tested by its ability to move its center of mass along a plane, and its success in doing so determines how much of its design goes into the next generation.

COMPUTER-BASED EVOLUTION OF VIRTUAL creatures was first demonstrated by Karl Sims in the early 1990s. Sims studied computer graphics at the MIT Media Lab, and Life Sciences as an undergraduate at MIT. He currently leads GenArts, Inc. in Cambridge, Massachusetts, which creates special effects software for the motion picture industry. His virtual creatures are often called Blockies, because they are 3-D assemblies of rectangular blocks of various sizes. (See http://www.biota.org/ksims/ for information and animations.)

A population of Blockies is created with a string of numbers, a "genotype," which is a code for a "phenotype" of blocks and connections. Each is tested to see how well it functions. Sims used four different types of tests, two assuming a "water" environment and two a "land" environment. Water Blockies were tested for how fast they could swim, or how well they could follow. Land Blockies were tested for how fast they could walk, or how high they could jump.

With each test, the genotypes of the most successful Blockies were combined, as in sexual reproduction to produce a new population of somewhat more successful Blockies. This process is carried on through hundreds of generations of graphical creatures.

GOLEMS—MAKING BLOCKIES A PHYSICAL REALITY

Blockies existed only on a computer, but in the late 1990s, a team at Brandeis University took a bolder step—the Golem Project. This linked up Blockie-style computer evolution to the type of rapid prototyping machine used in computer-aided manufacturing. These machines use a temperature-controlled head to extrude thermoplastic material layer by layer, eventually creating a solid, three-

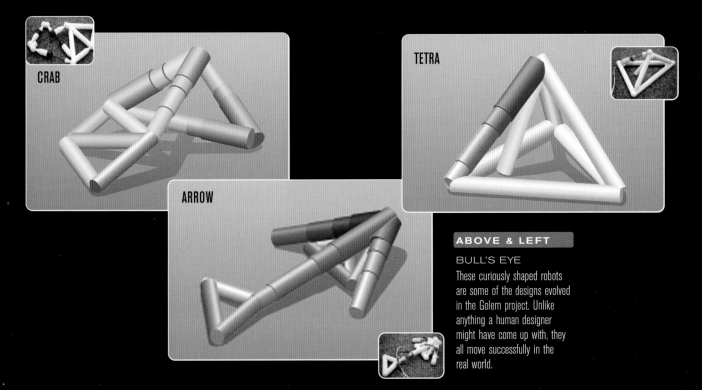

CRAB

ARROW

TETRA

ABOVE & LEFT

BULL'S EYE
These curiously shaped robots are some of the designs evolved in the Golem project. Unlike anything a human designer might have come up with, they all move successfully in the real world.

dimensional structure without human intervention.

The robots to be designed looked very much like Blockies—and very unlike any of the other robots in this book. They were made of bars and joints, and not only their morphology (shape) was to be evolved but also their control system, made of artificial neurons.

The evolution started with an initial population of 200 "null" robots with zero bars and zero neurons. Each robot was tested for locomotion ability, by seeing how far its center of mass moved across an infinite plane in a given time. As with the Blockies, the most successful designs were used to "breed" a new generation, by adding, modifying, and removing building blocks, and then replacing the old designs in the population with the new.

Tens of generations would pass before any actual movement occurred. To achieve this, a machine must at least manage to connect a bunch of artificial neurons

generating movement to a bit of body that could move. The process typically continued for between 300 and 600 generations before it produced successful robot designs. These were then downloaded to the rapid prototyping machine, which would construct the physical robot.

The motors and electronics still had to be added by hand, and the whole project still depended on computers and machinery built by humans. So the resulting robots were not truly self-replicating. But they were real robots, with real movement, and autonomously designed.

The team distributed their software to computer users round the world for use as a screensaver, so that idle PCs could be used for evolving more robots. Several million CPU (central processing unit) hours were used in total. But eventually the evolutionary process hit a complexity barrier. The team is now working on more powerful evolutionary programs.

BELOW

SMOOTH OPERATOR
This is the bar and joint mechanism generated by the prototyping machine.

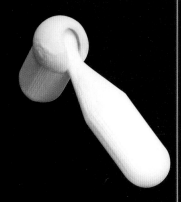

THERE'S NOTHING LIKE TRYING TO BUILD A ROBOT TO MAKE YOU APPRECIATE WHAT amazing creatures we human beings really are. And not just because of the achievements we normally think of as "intelligent."

Take human mobility: even walking is a sophisticated balancing act, but one we manage with ease. Running, jumping, and climbing are also straightforward for many humans—so our movement is not only complex but very flexible. Then there's hand-eye coordination and body awareness—most of us can touch our noses with our forefinger with our eyes shut, or reach out and grasp a coffee mug without knowing exactly where it is.

Humans are clever thinkers, but we also know when to ponder, and when to go ahead and act, which is just as important. We may put a lot of effort into learning a new skill, but once we understand how to do it, our new knowledge turns into a fast behavior. Good drivers are ones who think about where they are going, not how to operate the controls.

The amazing human visual system locates objects and identifies them from all kinds of angles and in different lighting conditions. We make the smooth link between behavior and thought without even realizing it. For adults at least, all this is often done without conscious effort. So it takes a huge leap of the imagination for people to understand just how hard it is to provide a robot with even a fraction of these abilities.

Humans can "think outside the box," solve novel problems, produce new and original ideas. We can imagine, dream, and create. Artificial intelligence is still only scratching the surface of all this. We've seen that robots can be given some learning abilities, but these are nowhere near comparable to those of a small child learning to communicate in language.

HUMANS RULE OK!

What a piece of work is a man!
how noble in reason! how infinite in faculties!
in form and moving how express and admirable!
in action how like an angel!
in apprehension how like a god!
the beauty of the world, the paragon of animals!

(*Hamlet*, in the Shakespeare play of the same name)

WHAT ABOUT THE ROBOTS?

In this book, we have seen important ideas investigated and fascinating achievements demonstrated. But tremendous effort is still needed to turn most of this work into robots that can become permanent inhabitants of the human world. In a real application—from cleaning to urban search and rescue—it is still basic engineering that currently carries the day, not amazing intelligence or lessons from biology.

Robotics is not only fascinating, it is also extremely hard work. The real world is much less forgiving than the inside of a computer, and much more challenging than a film studio. Robotics needs patience, not a headlong rush for instant success followed by disillusionment if it doesn't materialize.

Evolution has had millions of years to produce the living creatures—including humans—that are so well adapted to the world they inhabit. The dream of the living robot is only a few thousand years old, and the technology to begin serious work on realizing the dream only a fraction of that age. The more we understand of biology, the more clearly we see what a wonderful thing life is, and how far our robot designs are from matching it. But it works both ways—the more we try out our understanding of living things on robots, the more we come to understand the mechanics of life. And in this age of bioengineering, that understanding could turn out to be very important indeed.

ABOVE/TOP LEFT

INCOMPARABLE INTELLIGENCE
Human mobility and sensing, let alone human thought and creativity, are way ahead of what we can do with robots.

BOTTOM LEFT

TWENTIETH CENTURY MAN
The real world is much less forgiving than the film studios in which imagined robots like this are produced.

Making Your Own Robot

There's plenty of room in robotics for anyone who wants to have a go. So what are your options if you want to build a robot?

KITS

A search on the Internet will turn up dozens of robot kits. One of the best known is LEGO Mindstorms (http://mindstorms.lego.com/). With this assembly kit, the robot body is made of LEGO bricks; LEGO Technix motors and gears allow the robot to achieve mobility; and "smart bricks" containing microprocessors developed with the help of the MIT AI Lab provide the control system. The kit comes with a neat scripting language based on LOGO, the programing language developed by Seymour Papert of MIT in the 1970s, which makes it very easy to get started. Most Junior League RoboCup teams use it, and LEGO even run their own competitions. The LEGO kit has a very active user community—have a look at http://mindstorms.lego.com/community. This community has provided extra programing environments—such as NQC (Not Quite C), and the Forth programing language to replace the smart brick firmware. The kit is also widely used for teaching at all levels from high school to university.

The advantage of building robot bodies out of LEGO bricks is that it is easy to create many different structures and to alter the robot's shape. One disadvantage is that bodies are heavy and rather fragile—it's easy to knock bits off, especially if it is not programed in obstacle avoidance.

KIND OF A KIT?

Look at the BEAM (Biology, Electronics, Aesthetics, and Mechanics) idea at www.solarbotics.com. BEAM arises from the work of Mark Tilden of the Los Alamos National Laboratory in the United States (see www.nis.lanl.gov/projects/robot//html/contact.html). More of a philosophy than a kit, though supported by a set of components, BEAM emphasizes simple electronic circuitry, very often solar- or other light-powered. More advanced BEAM robots are battery powered. The solar-powered robots have no "off" switch and function even in poor light by storing what energy they collect in a capacitor and releasing it in short bursts to power their small, efficient motors.

Electronics are the core strength of BEAM robots, in the spirit of the Grey Walter robots. A BEAM robot supports insectlike behaviors, such as moving toward or away from a light source and avoiding obstacles with touch-sensitive antennas.

AND WITHOUT A KIT?

Imagination is the name of the game: you can build a robot without a kit using everyday items for body, sensors, and brain. There are many websites that can help, including The Robot Store (www.robotstore.com) and the BBC (www.bbc.co.uk/science/robots/techlab/). The Arrick Robotics website at www.robotics.com/robots.html is also very helpful.

Another way of making a robot body is to use Fisher-Technic, the German-designed construction kit, which is more solid than LEGO. Or you could use the body of a radio-controlled car, removing the radio control. Or the base of a child's stroller—anything mobile.

A flying robot could be based on a model plane or helicopter. A more manageable flying robot can be created under a helium balloon if you have access to a source of helium. You would then need some small motor-driven fans for propulsion and a way of angling them to change the robot's altitude.

A robot body needs sensors. You can easily get hold of these from electronics retailers or on the Internet. A cheap laser rangefinder can be produced by adapting an electronic tape measure. Webcams can function as an inexpensive, if rudimentary, vision system.

Adding control can be done from first principles by buying a controller board from the electronics shop or on the Internet. Another option is a Personal Digital Assistant (PDA) for smaller robots or an old laptop for larger ones.

HELPFUL BOOKS:

Build Your Own Robot by Karl Lunt of the Seattle Robotics Society
Robot Builder's Bonanza by Gordon McComb
Robot Building for Beginners by David Cook of Motorola

HIGH FASHION, HI-TECH?

This might look like someone wearing a kitchen implement on her head; in fact, it is a robot-controlled hat made at the MIT media lab. The robot in control is a LEGO smart brick known as a Cricket. One Cricket is attached to the hat; another is controlling an experiment bottom right, and sending data to the first Cricket via an infrared communication system. The results will be displayed on the dials of the hat.

HOBBYBOTS

GLOSSARY

active sensing – sensing the environment by bouncing a signal off it and analyzing the reflected signal.

actuator – a part of the robot which can be moved.

artificial neural net – a computer program built on nodes and connections using knowledge of how brains work.

attentional focus – focusing a sensor only on the "interesting" part of its input.

autopoiesis – the mechanisms that allow living creatures to maintain their processes and go on living.

behavior – a tightly linked package of sensing and acting with little or no processing in between.

biomechanics – the study of living things as mechanical structures.

biomimetics – the construction of artefacts to mimic biological creatures in some way.

boids – fictitious computer-based creatures used by Craig Reynolds in his original work on flocking.

degrees of freedom – the number of independent directions of motion a robot has to control.

echolocation – active sensing of the environment by bouncing sound off it; used by bats and dolphins.

ecosystem – a system of mutually-interacting, living creatures including all animals, insects, and plants.

g – a unit of measurement for the acceleration of gravity.

gait – the pattern formed by the motion of the legs on a creature with two or more legs.

genetic algorithms – computer software that mimics some of the processes by which evolution occurs.

genotype – the genetic endowment of a living entity.

homeostasis – a dynamic process that allows a system to keep itself in some constant state.

infrared – light just outside the visible red on the spectrum; often used for robot sensors.

kinematics – static analysis of how a robot actuator can get from one position to another by moving a joint.

localization – the process by which a robot works out where it is.

motion capture – the use of markers on the joints of a living thing (usually a human) to record its motion.

noise – the random variation in all electromagnetic signals, heard as a hiss; or actual interference, heard as clicks, etc.

odometer – a sensor that measures how many times a robot wheel has turned.

optic flow – the speed with which objects in the environment appear to move across the field of a sensor.

passive sensing – sensing the environment by receiving data it gives out.

perceptual map – a map of an environment constructed by linking together sensor readings taken from it.

phenotype – the physical animal that results from its genotype.

pheromone – a chemical message given out by animals; sometimes deposited in the environment.

phototaxis – behavior moving a robot to a light source.

piezoelectric – a material that generates a small amount of electricity when compressed.

pixel – short for "picture element": the smallest part of a visual field a sensor can resolve.

prioperception – the internal sensors that allow animals to know where their body parts are relative to each other.

rangefinder – an active sensor used to find the distance of objects in the environment.

reaction – an animal behavior tied closely to an incoming sensory stimulus.

reflex – a "hard–wired" behavior that must always result from a particular sensory stimulus.

reinforcement learning – computer software that uses rewards from successful behaviors to make them more likely to be used again.

sensor – a part of a robot that can receive data from its environment

serpentine motion – the wave-based motion of a snake.

stigmergy – communication via the environment, for example using pheromones.

stimulus – a significant sensory input, usually linked to some particular behavior or response.

taxis – the way an animal reacts to some kind of stimulus– a light or a sound for example–by moving determinedly either toward or away from it.

translation – robot movement in a straight line forward or backward.

ultrasound – high-pitched sound outside the human range; often used for robot sensors.

voxel – a 3-D pixel.

zero moment point – a dynamic center of gravity; the point on the robot where all forces are in balance.

MORE ABOUT ROBOTICS ON THE WEB:

The Catalogue of walking machines:
www.fzi.de/ipt/WMC/walking_machines_katalog/walking_machines_katalog.html

The World of Biomorphic Robotics:
www.iguana-robotics.com/RobotUniverse/BiomorphicRobots.htm

Internet Robotics Sources: www.cs.indiana.edu/robotics/world.html

The Robotics Institute at Carnegie-Mellon University: www.ri.cmu.edu/

Robotics at the MIT AI Lab: www.ai.mit.edu/people/brooks/projects.shtml

Books about robots: www.robotbooks.com/

FURTHER READING:

Cambrian Intelligence – The Early History of The New AI
Rodney Brooks, MIT Press; ISBN: 0262522632 (1999)

Flesh and Machines: How Robots Will Change Us
Rodney Brooks, Pantheon Books; ISBN: 0375420797 (2002)

An Introduction to AI Robotics
Robin Murphy, MIT Press; ISBN: 0262133830 (2000)

Emergence: The Connected Lives of Ants, Brains, Cities, and Software
Steven Johnson Scribner; ISBN: 068486875X (2001)

Digital Biology
Peter J. Bentley, Simon & Schuster; ISBN: 0743204476 (2002)

Artificial Minds
Stan Franklin, MIT Press; ISBN: 0262561093; Reprint edition (1997)

Creation: Life and How to Make It
Steve Grand, Harvard Univ Pr; ISBN: 0674006542 (2001)

The Society of Mind.
Marvin L. Minsky, Simon & Schuster (Paper); ISBN: 0671657135 (1988)

Vehicles
Valentino Braitenberg, MIT Press; ISBN: 0262521121 (1986)

River Out of Eden: A Darwinian View of Life
Richard Dawkins, (Science Masters Series) Basic Books; ISBN: 0465069908

The Darwin Wars
Andrew Brown, Pocket Books; ISBN: 0684851458 (2000)

ACKNOWLEDGMENTS

So many people helped so much in the hard labor of producing this book, it's difficult to list them all. If I have left anyone out, put it down to all the chaos of trying to write a book while doing all the other things university people are supposed to do, not to any lack of gratitude. From robotics researchers I have never met, to friends who dug out special photos, thank you all, including: David Barnes, University of Wales, Aberystwyth, UK; David Beal, Massachusetts Institute of Technology, US; Aude Billard, University of Southern California, US; Ian Briggs and family, UK; Sharon Campbell, iRobot, US; Jeffrey Cohn, Carnegie-Mellon University, US; Mark Colton, University of Utah, US; Hanna and Heather Craig, US; Holk Cruse, University of Bielefeld, Germany; Kerstin Dautenhahn, University of Hertfordshire, UK; James DeLaurier, University of Toronto, Canada; Steve DeWeerth, Georgia Institute of Technology, US; Fabio Di Francesco, CNR, Italy; Jonas Dino, NASA, US; Max Petre Eastty and Grace Petre Eastty, UK; Hugo Elias, Shadow Robot Company, UK; Nancy Garcia, Sandia National Laboratories, US; James Garry, Open University, UK; Birgit Graf, Fraunhofer IPA, Germany; Steve Grand, Cyberlife Research, UK; Mike Hamilton, AAAI Publications, US; Richard Hearne, Open University, UK; James Hendler, University of Maryland, US; Jesse Hoey, University of British Columbia, Canada; Owen Holland, University of Essex, UK; Huosheng Hu, University of Essex, UK; Greg Lock, US; Peter Kahn, University of Washington, US; Ian Kelly, California Institute of Technology, US; Graham Kirton, UK; Glenn Klute, University of Washington, US; Hiroshi Kobayashi, Science University of Tokyo, Japan; Marianne LaFrance, Yale University, US; Maja Mataric, University of Southern California, US; Yoky Matsuoka, Carnegie-Mellon University, US; Jim McCann, BBH Inc, US; Iain Matthews, Carnegie-Mellon University, US; Francois Michaud, Sherbooke University, Canada; Robert Michelson, Georgia Institute of Technology, US; Jamie Montemayor, University of Maryland, US; Steve Musgrave and family, UK; Ian Mudie, University of West of England, UK; Monica Nicolescu, University of Southern California, US; Simon Penny, University of California, Irvin, US; Marion Petre, Open University, UK; Annika Pfluger, Massachusetts Institute of Technology, US; Giovanni Pioggia, University of Pisa, Italy; John Pursley, Envision Product Design LLC, US; Roger Quinn, Case Western Reserve University, US; Andy Ruina, Cornell University, US; Noel Sharkey, Magna and University of Sheffield, UK; Sarah Shrive, University of Luton Press, UK; Elizabeth Sklar, Columbia University, US; Luc Steels, VUB Belgium and Sony Computer Lab, Paris, France; David Touretzky, Carnegie-Mellon University, US; Michael Triantafyllou, Massachusetts Institute of Technology, US; Alexander Van de Rostyne, Belgium; Rodrigo Ventura, Instituto Superior Tecnico, Lisboa, Portugal; Gurvinder Virk, University of Portsmouth, UK; Anne Watzman, Carnegie-Mellon University, US; Alik Widge, Carnegie-Mellon University, US; Stuart Wilkinson, University of South Florida, US.

INDEX

Figures in italics refer to captions.

CREDITS

Quarto would like to thank and acknowledge the following for images reproduced in this book:

Key: t = top, b = bottom, l = left, r = right, c = center,

7 Topham; 9 l Topham; 10 Topham; 11 t Ann Ronan, b Wax Museum, Prague; 12 l Ann Ronan, r Wax Museum, Prague; 13 Ann Ronan; 14 t Topham, b Topham; 15 c Topham, t Ann Ronan; 16 Ann Ronan; 18 TRIP; 21 Honda Motor Co.; 22 tl & tr Robot-Assisted Search and Rescue (CRASAR); 23 tl Robot-Assisted Search and Rescue (CRASAR), tr Topham, br The Planetary Society; 24 Topham; 25 cr & bc Roger Quinn, Case Western Reserve University; 27 tl Topham, cr Cornell University; 29 cl Gurvinder Singh, University of Portsmouth, br iRobot; 30 Nasa, photographer Dominic Hart; 30 t and 31 b Erich Rome (c) 2000 GMD; 32 tl & cl MIT Deptartment of Ocean Engineering; 33 bl MIT Deptartment of Ocean Engineering, b TRIP, cr iRobot; 34 tr Alexander Van de Rostyne; 35 tl & tr James DeLaurier, University of Toronto, and James F. Winfield; 36 and 37 Steve Grand 39 l University of Bath; 39 r Optronix; 40 tl & tr Graham Kirton; 42 bl & 43 tr (c) Owen Holland; 43 tl Topham/Image Works; 44 bl Greg Locke, tr Ian & Sue Briggs; 45 Graham Kirton; 49 br MIT AI laboratory; 50 b TRIP, t Joseph Ayers, NorthEastern University, Marine Science Centre; 51 tr Universtiy of Pisa, br University of Portsmouth; 53 University of the West of England; 54 NASA and Carnegie-Mellon University; 55 U.S. Department of Defense; 57 TRIP; 60 Funio Hara and Hiroshi Kobayashi, Science University, Tokyo; 61 Science Photo Library; 63 TRIP; 66 t Topham; 66 b Topham; 67 Barbara Webb, Stirling University; 69 Steve DeWeerth, Georgia Institute of Technology; 71 David Barnes, University of Aberystwyth; 72 t Maja Mataric, University of South Carolina; 74 bl David Barnes, Universtiy of Aberystwyth; 77 t Sandia Labs, photographer: Randy Montoya, b NASA; 78, 79 Dr. Huosheng Huo and the Intelligent Inhabited Environments group, University of Essex; 82 Ann Ronan; 83 tr Dr. Huosheng Huo and the Intelligent Inhabited Environments group, University of Essex; 84 Max and Grace Petre; 87 cb Steve Musgrave, r Empics; 88 Institute of Systematic Coaching and Training; 90 tl Shadow Robot group, United Kingdom, tr Roger Quinn, Case Western Reserve University; 91 Yoky Matsuoka, Carnegie Mellon University; 93 tr NASA, bl Hiroshi Kobayashi, University of Tokyo; 95 tr Sandia Laboratories, bl François Michaud; 96 bl Robert Michelson; 97 tl and tr Stuart Wilkinson; 98 Sandia Laboratories; 99 Ann Ronan/Twentieth Century Fox, 102/103 Sam Ogden/Science Photo Library; 105 Peter Kahn and Batya Friedman, University of Washington, and Alan Beck and Nancy Edwards, Purdue University; 107 tl and tc Cynthia Brezeal, MIT, br Dolores Canamero, University of Hertfordshire; 108 Marianne LaFrance, Yale University; 111 Hiroshi Kobayashi, University of Tokyo; 112 tc (inset) Peter Kahn and Batya Friedman, Washington University, and Alan Beck and Gail Melson, Purdue University; 113 Aude Billard, University of Southern California; 114 t Museumsstiftung Post und Kommunikation, Berlin, l Kerstin Dautenhahn, University of Herfordshire; 115 Fraunhofer IPR; 116 r Department of Computer Science, University of Maryland; 117 Peter Kahn and Batya Friedman, University of Washington, and Alan Beck and Gail Melson, Purdue University; 121 Professor Noel Sharkey; 123 t Topham; 124 l Trip; 125 l Empics; 126 courtesy of Electrolux; 127 inset Topham, 128 Professor Noel Sharkey; 129 t AI Lab, Vrije Universiteit Brussels; 130 Intelligent Autonomous Systems Laboratory, University of the West of England; 131 Intelligent Autonomous Systems Laboratory, University of the West of England; 133 Computer Science Dept, Brandeis University, 134 b Ann Ronan; 137 Sam Ogden/Science Photo Library

All other illustrations are the copyright of Quarto. While every effort has been made to credit contributors, we apologize should there have been any omissions or errors.